Lecture Notes in Computer Science 16193

Founding Editors

Gerhard Goos
Juris Hartmanis

Editorial Board Members

Elisa Bertino, *Purdue University, West Lafayette, IN, USA*
Wen Gao, *Peking University, Beijing, China*
Bernhard Steffen , *TU Dortmund University, Dortmund, Germany*
Moti Yung , *Columbia University, New York, NY, USA*

The series Lecture Notes in Computer Science (LNCS), including its subseries Lecture Notes in Artificial Intelligence (LNAI) and Lecture Notes in Bioinformatics (LNBI), has established itself as a medium for the publication of new developments in computer science and information technology research, teaching, and education.

LNCS enjoys close cooperation with the computer science R & D community, the series counts many renowned academics among its volume editors and paper authors, and collaborates with prestigious societies. Its mission is to serve this international community by providing an invaluable service, mainly focused on the publication of conference and workshop proceedings and postproceedings. LNCS commenced publication in 1973.

Lei Li · Viktor Jirsa · Jianfeng Feng · Jun Deng ·
Luca Dede' · Sora An · Yilin Lyu · Xiaoyue Liu
Editors

Digital Twin for Healthcare

First International Workshop, DT4H 2025
Held in Conjunction with MICCAI 2025
Daejeon, South Korea, September 23, 2025
Proceedings

 Springer

Editors
Lei Li
National University of Singapore
Singapore, Singapore

Viktor Jirsa
Aix-Marseille Université
Marseille, France

Jianfeng Feng
Fudan University
Shanghai, China

Jun Deng
Yale University
New Haven, CT, USA

Luca Dede'
Politecnico di Milano
Milan, Milano, Italy

Sora An
Ewha Womans University
Seoul, Korea (Republic of)

Yilin Lyu
National University of Singapore
Singapore, Singapore

Xiaoyue Liu
National University of Singapore
Singapore, Singapore

ISSN 0302-9743 ISSN 1611-3349 (electronic)
Lecture Notes in Computer Science
ISBN 978-3-032-07693-9 ISBN 978-3-032-07694-6 (eBook)
https://doi.org/10.1007/978-3-032-07694-6

© The Editor(s) (if applicable) and The Author(s), under exclusive license to Springer Nature Switzerland AG 2026

This work is subject to copyright. All rights are solely and exclusively licensed by the Publisher, whether the whole or part of the material is concerned, specifically the rights of translation, reprinting, reuse of illustrations, recitation, broadcasting, reproduction on microfilms or in any other physical way, and transmission or information storage and retrieval, electronic adaptation, computer software, or by similar or dissimilar methodology now known or hereafter developed.
The use of general descriptive names, registered names, trademarks, service marks, etc. in this publication does not imply, even in the absence of a specific statement, that such names are exempt from the relevant protective laws and regulations and therefore free for general use.
The publisher, the authors and the editors are safe to assume that the advice and information in this book are believed to be true and accurate at the date of publication. Neither the publisher nor the authors or the editors give a warranty, expressed or implied, with respect to the material contained herein or for any errors or omissions that may have been made. The publisher remains neutral with regard to jurisdictional claims in published maps and institutional affiliations.

This Springer imprint is published by the registered company Springer Nature Switzerland AG
The registered company address is: Gewerbestrasse 11, 6330 Cham, Switzerland

If disposing of this product, please recycle the paper.

Preface

Digital twin (DT) technology is emerging as one of the most promising frontiers in healthcare. At its core, it enables the creation of dynamic, patient-specific virtual models that integrate medical imaging, physiological data, and computational simulations to mirror and predict real-world health outcomes. Unlike static models, DTs can evolve alongside the patient, continuously incorporating updates from imaging, sensors, and clinical records to reflect changes in anatomy, physiology, and disease progression. Such adaptability offers a safe and controllable virtual environment in which hypotheses can be tested, treatments simulated, and clinical workflows optimized before they are applied in practice.

Within the MICCAI community, DT research is still in its early stages but represents a natural extension of the field's strengths in medical image analysis, computational modeling, and AI-driven decision support. Traditional image computing has focused on one-time, static representations of anatomy or function. DTs add a longitudinal and adaptive dimension, enabling predictive and even prescriptive healthcare strategies. This paradigm bridges multiple disciplines, such as computational physiology, biomechanics, trustworthy and explainable AI, and in silico clinical trials. It opens new avenues for both methodological innovation and clinical translation. These advances are timely, as healthcare systems worldwide face growing demands for precision, efficiency, and equitable access to advanced diagnostics and treatments. By broadening MICCAI's scope, DTs also have the potential to support real-time patient monitoring, enable population-scale imaging informatics, and deliver accessible solutions tailored to underrepresented communities. An overview of these objectives is available on the workshop website: https://digitaltwinforhealthcare.com/.

The Digital Twin for Healthcare 2025 (DT4H 2025) workshop was held on September 23, 2025, in Daejeon, South Korea, in conjunction with MICCAI 2025. The workshop was designed to foster discussion on integrating DT frameworks with conventional medical imaging and computational tools, and to build a new DT-focused sub-community within MICCAI. The program covered key topics such as organ- and disease-specific DT modeling; multi-modal data integration (imaging, signals, genomics, biosensors); AI-driven simulation and prediction; DT-guided diagnostic and treatment planning; verification, validation, and uncertainty quantification; scalable/federated learning; and ethical, regulatory, and translational challenges.

A total of 24 qualified submissions were received from a range of countries, including China, Singapore, the UK, Switzerland, South Korea, the USA, Germany, and France. All submissions underwent a double-blind peer review, following the MICCAI 2025 reviewer guidelines. Each paper was anonymously evaluated by an average of three independent reviewers, with each reviewer assigned no more than three papers. Reviewers assessed submissions based on novelty, methodological rigor, clinical relevance, and clarity. The full review reports and evaluation criteria are available on the workshop website for readers' reference. Following peer review and program committee deliberation, 15 full papers were accepted, yielding an acceptance rate of 65%. There were no invited papers included in these proceedings.

We extend our sincere gratitude to the program committee members, external reviewers, and all authors for their valuable contributions. We also thank our sponsors, including Amazon Web Services, BAAI, SIG-Cardiac, and Digital Heart Lab., for their support. We hope that this volume will serve as a foundation for further exploration and collaboration in digital twin applications in healthcare.

November 2025

Lei Li
Viktor Jirsa
Jianfeng Feng
Jun Deng
Luca Dede'
Sora An
Yilin Lyu
Xiaoyue Liu

Organization

Program Committee Chairs

Lei Li	National University of Singapore, Singapore
Viktor Jirsa	Institut de Neurosciences des Systèmes, France
Jianfeng Feng	Fudan University, China
Jun Deng	Yale University, USA
Luca Dede'	Politecnico di Milano, Italy
Sora An	Ewha Womans University, South Korea

Program Committee

Yilin Lyu	National University of Singapore, Singapore
Xiaoyue Liu	National University of Singapore, Singapore

Additional Reviewers

Alexander Zolotarev
Chaoran E.
Encheng Su
Fan Yang
Hao Li
Jihe Li
Julia Camps
Lei Liu
Linwei Wang
Luca Dede
Qingya Li
Shijie Wang
Sora An
Xiaohan Yuan
Xiaoxian Zhang
Xiaoyue Liu
Xuan Yang
Yilin Lyu
Yinling Zhu

Contents

Personalized 3D Myocardial Infarct Geometry Reconstruction from Cine
MRI with Explicit Cardiac Motion Modeling 1
   Yilin Lyu, Fan Yang, Xiaoyue Liu, Zichen Jiang, Joshua Dillon,
   Debbie Zhao, Martyn Nash, Charlene Mauger, Alistair Young,
   Ching-Hui Sia, Mark Y. Chan, and Lei Li

Microvascular Retinal Digital Twins from Non-invasive Clinical Images 12
   Rémi J. Hernandez and Wahbi K. El-Bouri

Validating Digital Twins with Tactile-Visual Liver Phantoms
for Robot-Assisted Surgical Workflows 23
   Chengzheng Mao, Ying Zhen Tan, and Yujia Gao

A Real-Time Digital Twin for Type 1 Diabetes Using Simulation-Based
Inference 35
   Trung-Dung Hoang, Alceu Bissoto, Vihangkumar V. Naik,
   Tim Flühmann, Artemii Shlychkov, Jose Garcia-Tirado, and Lisa M. Koch

Retrospective Evaluation of a Patient-Specific Liver Digital Twin
to Predict Thermal Ablation Outcomes in HCC 47
   Chloé Audigier, Felix Meister, Fouad Georges Akkari, Andrea Tonglet,
   Oliver Frings, and Rafael Duran

Acoustic Simulation with Deep Learning for Low-Intensity Transcranial
Focused Ultrasound Digital Twins 58
   Minjee Seo, Minwoo Shin, Gunwoo Noh, Seung-Schik Yoo,
   and Kyungho Yoon

Towards Digital Twin of RF Ablation: Real-Time Prediction
of Time-Dependent Thermal Effects Using Transformer 69
   Seonaeng Cho, Minjee Seo, Minwoo Shin, and Kyungho Yoon

Finite-Element Electrophysiological Modeling of Human Uterine Smooth
Muscle Using a Reduced Tong Model 79
   Zhen Li and Alberto Corrias

TF-TransUNet1D: Time-Frequency Guided Transformer U-Net for Robust
ECG Denoising in Digital Twin 90
   Shijie Wang and Lei Li

DeformMLP: Effective Deformation Prediction for Breast Cancer Using
Graph Topology-Assisted MLPs 99
   *Yong-Min Shin, Kyunghyun Lee, Sunghwan Lim, Kyungho Yoon,
and Won-Yong Shin*

Rule-based Key-Point Extraction for MR-Guided Biomechanical Digital
Twins of the Spine 109
   *Robert Graf, Tanja Lerchl, Kati Nispel, Hendrik Möller, Matan Atad,
Julian McGinnis, Julius Maria Watrinet, Johannes Paetzold,
Daniel Rueckert, and Jan S. Kirschke*

Towards Robust Algorithms for Surgical Phase Recognition via Digital
Twin Representation 119
   *Hao Ding, Yuqian Zhang, Wenzheng Cheng, Xinyu Wang, Xu Lian,
Chenhao Yu, Hongchao Shu, Ji Woong Kim, Axel Krieger,
and Mathias Unberath*

Personalized 4D Whole Heart Geometry Reconstruction from Cine MRI
for Cardiac Digital Twins 130
   *Xiaoyue Liu, Xicheng Sheng, Xiahai Zhuang, Vicente Grau,
Mark YY Chan, Ching-Hui Sia, and Lei Li*

Secure Medical Digital Twins: A Use-Case Driven Approach 140
   *Salmah Ahmad, Bianca Bartelt, Matthias Enzmann, Jörn Kohlhammer,
Stefan Wesarg, and Ruben Wolf*

Explainable Prediction of Recurrence After Prostate Cancer Radiotherapy
Using *in Silico* digital twin model and machine learning 152
   *Valentin Septiers, Carlos Sosa-Marrero, Eleonora Poeta,
Hilda Chourak, Aurélien Briens, Renaud De Crevoisier,
Maria A. Zuluaga, and Oscar Acosta*

Author Index 165

Personalized 3D Myocardial Infarct Geometry Reconstruction from Cine MRI with Explicit Cardiac Motion Modeling

Yilin Lyu[1], Fan Yang[1], Xiaoyue Liu[1], Zichen Jiang[2], Joshua Dillon[3], Debbie Zhao[3], Martyn Nash[3], Charlene Mauger[4], Alistair Young[4], Ching-Hui Sia[5,6], Mark Y. Chan[5,6], and Lei Li[1(✉)]

[1] Department of Biomedical Engineering, National University of Singapore, Singapore, Singapore
lei.li@nus.edu.sg
[2] Department of Computer Science, National University of Singapore, Singapore, Singapore
[3] Auckland Bioengineering Institute, University of Auckland, Auckland, New Zealand
[4] School of Biomedical Engineering and Imaging Sciences, King's College London, London, UK
[5] Department of Medicine, National University of Singapore, Singapore, Singapore
[6] Department of Cardiology, National University Heart Centre Singapore, Singapore, Singapore

Abstract. Accurate representation of myocardial infarct geometry is crucial for patient-specific cardiac modeling in MI patients. While Late gadolinium enhancement (LGE) MRI is the clinical gold standard for infarct detection, it requires contrast agents, introducing side effects and patient discomfort. Moreover, infarct reconstruction from LGE often relies on sparsely sampled 2D slices, limiting spatial resolution and accuracy. In this work, we propose a novel framework for automatically reconstructing high-fidelity 3D myocardial infarct geometry from 2D clinically standard cine MRI, eliminating the need for contrast agents. Specifically, we first reconstruct the 4D biventricular mesh from multi-view cine MRIs via an automatic deep shape fitting model, biv-me. Then, we design an infarction reconstruction model, CMotion2Infarct-Net, to explicitly utilize the motion patterns within this dynamic geometry to localize infarct regions. Evaluated on 205 cine MRI scans from 126 MI patients, our method achieves a Dice score of 0.652 when compared with manual delineation, demonstrating reasonable agreement. This study demonstrates the feasibility of contrast-free, cardiac motion-driven 3D infarct reconstruction, paving the way for efficient digital twin of MI.

Keywords: Myocardial Infarction · Cine MRI · 3D Infarct Reconstruction · Cardiac Motion · Contrast Free

1 Introduction

Myocardial infarction (MI) remains a major cause of mortality and disability worldwide [16]. Accurately reconstructing the 3D geometry of infarcted regions is important for evaluating infarct size, extent, and location, which are clinically relevant for diagnosis, prognosis, and therapy planning. Recently, computational modeling of patient-specific hearts has emerged as a promising noninvasive tool for guiding personalized treatment [6,21]. Accurately representing myocardial remodeling in ischemic cardiomyopathy requires integrating patient-specific infarct geometry into these models [10]. Among clinical imaging techniques for infarct characterization, late-gadolinium enhanced magnetic resonance image (LGE MRI) is the most widely used [11,12]. While effective, LGE MRI requires contrast agent injection, which may cause side effects, increase scanning time, and reduce patient comfort [15]. In contrast, cine MRI, a standard clinical tool for visualizing cardiac anatomy and motion, offers non-invasive imaging of the heart without contrast agents. However, both LGE and cine MRI typically capture sparse, intersecting 2D planes, i.e., short-axis (SAX) and a few long-axis (LAX) slices, limiting spatial resolution and hindering the reconstruction of a detailed 3D heart model.

For 3D heart model reconstruction from 2D cardiac planes, many work employ two-stage model, i.e., image segmentation and 3D geometry reconstruction [2,4,9,26,28]. However, all these work only can reconstruct the 3D (or 3D + t) geometry model, where the infarction area is not identified. The computational modeling needs 3D infarct geometry, which is still coarsely estimated based on scar interpolation from LGE MRI [20]. Recently, several studies have explored scar analysis using contrast-free imaging as a more cost-effective alternative [23,24,29]. One widely adopted approach leveraged generative models to synthesize LGE-style images, enabling LGE-based analyses without the need for contrast agents. For example, Xu et al. [22] introduced sequential causal generative models that integrated synthesis and scar segmentation within an adversarial learning framework. However, these methods are highly dependent on the quality of the synthesized LGE images. An alternative strategy is to analyze cine MRI directly, utilizing its temporal information to detect abnormalities [23,25,29]. However, all these works only capture the cardiac motion information in the 2D single view. Liu et al. [13] proposed a 4D spatiotemporal framework, but the point clouds are primarily generated from single cine MRI (short-axis) views. Moreover, most current popular solutions only implicitly extract cardiac motion such as using optical flow-based methods, which primarily capture motion between adjacent frames.

In this study, we develop a 3D infarct geometry reconstruction model that leverages cardiac motion features extracted from multi-view cine MRIs. The proposed framework explicitly integrates cardiac morphology and motion dynamics to establish a relationship between abnormal myocardial motion and infarcted regions. This is accomplished by introducing a 4D cardiac mesh, derived from cine MRI, as the input to a cardiac motion mapping model (CMotion2Infarct-Net) for infarction localization. Furthermore, by leveraging the spatial

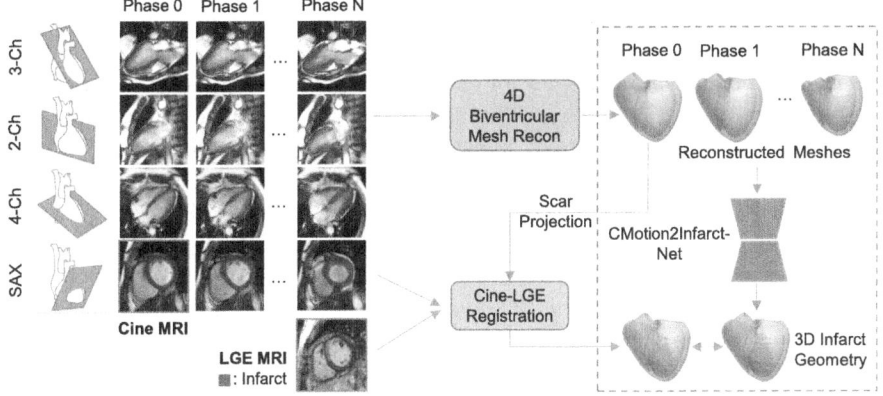

Fig. 1. Illustration of the multi-view cine-MRI based 3D infarct reconstruction framework. Note that the figure only presents a single short-axis (SAX) slice as an example, even though a stack of SAX view are employed here. LGE MRI is registered to the SAX end-diastolic (ED) phase, i.e., phase 0.

correspondence between cine and LGE MRI, we project the 2D scar information from LGE MRI onto the reconstructed cardiac mesh, providing supervision for CMotion2Infarct-Net training. To the best of our knowledge, this is the first study to directly reconstruct a structured 3D myocardial infarct geometry from cine MRI.

2 Methodology

Figure 1 provides an overview of the proposed 3D infarct geometry reconstruction model, consisting of 4D biventricular mesh reconstruction module, cine-LGE registration module, and cardiac motion mapping to infarct model. The 4D mesh reconstruction module integrates four different views of cine MRI together and fits them to a mesh template for cardiac reconstruction (Sect. 2.1). To train the CMotion2Infarct-Net, we generate the 3D Ground Truth (GT) infarct model as supervision based on cine and LGE registration and 3D scar projection (Sect. 2.2). Finally, Sect. 2.3 presents the details of the reconstruction model for the personalized inference of 3D infarct model.

2.1 4D Biventricular Model Reconstruction from Cine MRI

We adopt biv-me [7], an open-source and fully automated reconstruction pipeline, to infer 4D biventricular meshes from multi-view cine MRIs. It consists of three stages: view selection, segmentation, and cardiac geometric fitting. ResNet50 is first employed to identify and classify useful views within the cine MRI sequences. Next, the nnU-Net is used to segment the biventricular region, i.e., left ventricle (LV) cavity, right ventricle (RV) cavity, LV myocardium, and extract corresponding 2D contours from the selected views. Finally, these sparse contour sets are merged together based on their world coordinate and used to reconstruct biventricular meshes for each time frame through an iterative diffeomorphic registration algorithm. This is achieved by decomposing the deformations into two steps to ensure a bijective transformation. Specifically, a multi-class surface template mesh is first aligned to each contour set using an implicit linear least squares fit. The successive least-square fits can accelerate convergence and improve initialization at a lower computational cost. To preserve topology, the displacement of the coarse mesh is constrained within each iteration, ensuring that the Jacobian determinant remains positive. The explicit diffeomorphic fit further refines the alignment, ensuring a structured, point-correspondent mesh representation across the cardiac cycle.

2.2 Registration of Cine and LGE MRI for 3D Scar Projection

To generate a 3D representation of the infarct region for supervision, we leverage manual scar segmentation from LGE MRI and project the identified scars onto a 3D biventricular surface mesh reconstructed from cine MRI. Due to differences in spatial resolution, field of view, and respiratory motion, cine and LGE MRI are often misaligned. To address this, we employ a multivariate mixture model-based registration framework [30] to align cine and LGE MRI. The registration process involves identifying corresponding slices along the Z-axis, followed by in-plane rigid and non-rigid transformations. Once LGE MRI is spatially aligned with cine MRI, the LGE-derived annotations can be accurately mapped onto end-diastolic (ED) phase cine images. Nonetheless, the transformed infarct regions remain sparse due to the inherent slice thickness and limited coverage of 2D MRI acquisitions. Accordingly, we employ Gaussian sampling to generate denser scar distribution. Specifically, for each scar voxel identified in cine MRI, additional points are synthesized along the Z-axis by sampling from a normal distribution $\mathcal{N}(\mu, \sigma^2)$, where μ is the original Z-coordinate and $\sigma = 3$ mm. These augmented scar points are then mapped onto the 5 nearest vertices of the 3D heart surface mesh using a KDTree-based nearest-neighbor search. Note that for simplification in this study we project all scars onto LV endocardium, as used in Codreanu et al. [5]. Subsequently, the corresponding mesh vertices are labeled as scarred regions, ensuring a more spatially coherent and anatomically realistic infarct representation on the 3D surface mesh (Fig. 2).

2.3 Explicit Cardiac Motion Based Infarct Reconstruction

Given a sequence of 4D biventricular surface meshes, $\{M_t\}_{t=1}^{N}$, their spatial and motion characteristics can be captured using CMotion2Infarct-Net. CMotion2Infarct-Net comprises a preprocessing module, a spatio-temporal feature extraction module, and an attention based segmentation module. The module takes the biventricular model as input, while the network focus is on the LV myocardium. Accordingly, we extracted the endocardial mesh of LV, consisting of 1572 nodes, to serve as the input for the subsequent step. It is well recognized

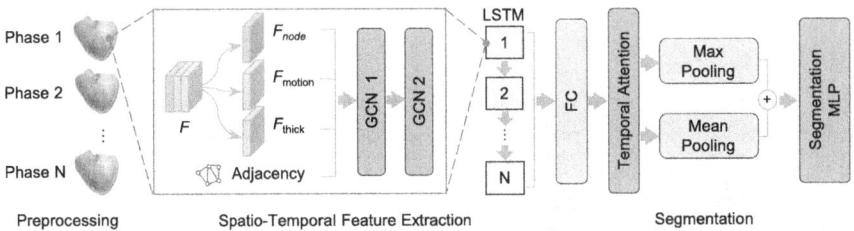

Fig. 2. The architecture of CMotion2Infarct-Net. The initial input was based on LV endocardial surface nodes, edges, and labels. N denotes the index of the phrase, F represents the features after preprocessing, F_{node}, F_{motion}, F_{thick} correspond to the features representing position, inter-phrase dynamics, and wall thickness, respectively.

that an important feature of MI is the abnormal motion [1,3,8,19]. Therefore, to further enhance the extraction of local features, we introduced first-order interphase differences as the local motion feature F_{motion}. Moreover, radial direction thickening is a major component of myocardial strain [17], and there exists a certain correlation between infarcted regions and wall thickening [14,27]. Therefore, we introduced the distance from the endocardium to the epicardium of LV to accurately compute the thickness F_{thick} as the initial inputs. The spatio-temporal feature extraction module first utilizes graph neural network (GNN) to extract structural features. After that, a two-layer long short-term memory (LSTM) network is applied to capture temporal dependencies across all phases, enabling point-wise feature extraction that integrates both spatial and temporal information. The segmentation head consists of a fully connected (FC) layer, a temporal attention layer, two spatial pooling operations (max pooling and mean pooling), and a MLP. Here, we design a transformer with 4 attention head to capture long dependencies and complex temporal interactions. Then, we utilize max pooling and mean pooling to extract global spatial features and concatenate for the next step. And finally, a two-layer segmentation MLP maps the latent features to the final segmentation output. CMotion2Infarct-Net is optimized by minimizing the regularized mesh segmentation loss, formulated as:

$$\mathcal{L}(M_{infarct}, \hat{M}_{infarct}) = \mathcal{L}_{BCEweighted} + \lambda_{Tversky}\mathcal{L}_{Tversky}(\alpha, \beta) \quad (1)$$

where $M_{infarct}$ and $\hat{M}_{infarct}$ represent the predicted and GT infarct, respectively. Since the infarct region typically occupies smaller area compared to the normal LV, $\mathcal{L}_{BCE}$, specifically a weighted binary cross entropy (BCE) loss is employed as the primary loss. $\mathcal{L}_{Tversky}$ represents the Tversky Loss, and α, β are weighting parameters to balance false positives and false negatives, respectively. λ are balancing parameters.

3 Experiments and Results

3.1 Materials

Data Acquisition. We collected 205 paired LGE and multi-view cine MRI scans from 126 post-MI patients at the National University Health System, Singapore. Each patient has up to two sets of scans, with a time interval of approximately 6 months to 12 months. During this period, the scar morphology may change. Specifically, a stack of SAX balanced steady-state free precession (bSSFP) cine MRI and three LAX cine sequences (2-, 3-, and 4-chamber views) have been acquired, as shown in Fig. 1. The SAX cine sequences consist of 8 to 17 slices across 25 frames. The dataset was randomly divided into 150 for training, 10 for validation, and 45 for test. To prevent data leakage, each patient's data is included in only one dataset. As a result, we use data from 93 patients for training, 6 for validation and 27 for test, respectively.

Implementation. The framework was implemented in Pytorch, running on a workstation equipped with an AMD EPYC 7K62 Processor and a NVIDIA GeForce RTX 4090 GPU. We use the Adam optimizer to update the network parameters via stochastic gradient decent (weight decay = 1×10^{-4}). The initial learning rate is set to 1×10^{-4} and multiplied by 0.7 every approximately 800 iterations. The parameters in Sect. 2.3 are set as follows: $\lambda_{Tversky} = 1$, and $\alpha=0.3$, $\beta=0.7$ to penalize false negatives more. The biv-me took about 9 min for each subject. CMotion2Infarct-Net was trained in 2 h (600 epochs), with an inference time of 5 s per case.

Gold Standard and Evaluation. To evaluate the reconstruction accuracy of 4D mesh, we manually segmented the biventricular area (LV, RV, and LV Myo) using ITK-SNAP on the cine data at the ED phase. We also calculated time-resolved chamber volumes for each reconstructed 4D heart. For 3D infarct evaluation, LGE MRIs were manually segmented by a trained student and verified by an expert. Ground-truth infarct regions on cine MRI were obtained by registering LGE segmentation to ED-phase cine images. In this study, the 3D infarct geometry of LV endocardium (generated in Sect. 2.2) is used as the GT. We then employed Dice score, Recall, ASD and Generalized Dice (G Dice) to assess the infarct region overlap and alignment between the predicted infarct geometry and the GT.

3.2 Results

Accuracy of 4D Biventricular Reconstruction. The average ASD between manually segmented contours and the 4D meshes reconstructed by biv-me is 2.55 ± 0.452 mm across 49 randomly selected subjects. Figure 3 (a) presents the overlap visualization of the biv-me predicted meshes and GT for four representative subjects. The predicted meshes are closely aligned with the sparse contours. To further assess motion accuracy, we analyzed volume change curves of the reconstructed 4D meshes. Figure 3 (b) illustrates two examples of the predicted heart shapes alongside the corresponding chamber volume variations over time. The results indicate that biv-me effectively captures the systolic and diastolic phases observed in cine MRI, highlighting its ability to preserve physiological motion dynamics. The alignment in shape and motion indicates accurate extraction of physiological deformation.

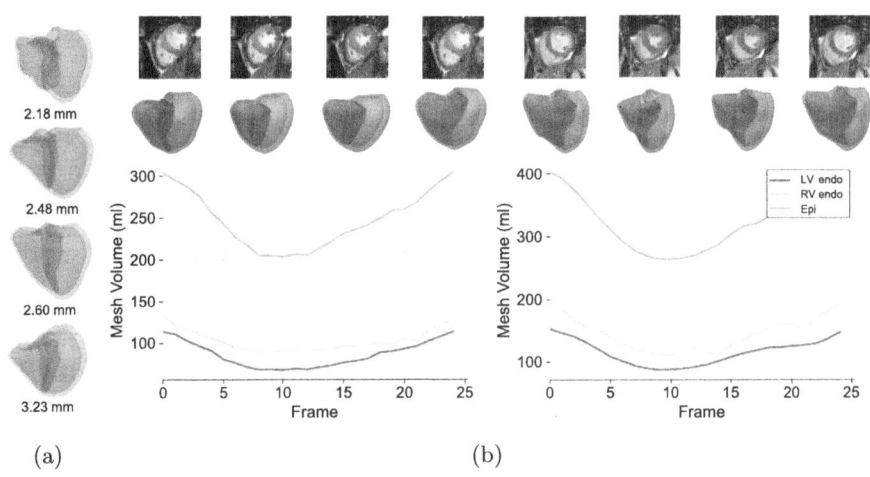

Fig. 3. (a) 3D visualization of the overlap between the sparse contours and reconstructed heart; (b) Illustration of cine MRI in short-axis view and the predicted 4D biventricular mesh with corresponding volume change over time.

Table 1. Summary of the quantitative evaluation results of 3D infarct reconstruction.

Method	Dice	Recall	ASD (mm)	G Dice
Inter-observer variation	0.798 ± 0.087	0.770 ± 0.125	1.158 ± 0.550	0.824 ± 0.079
CMotion2Infarct-Net	**0.652** ± 0.174	0.805 ± 0.179	**3.135** ± 2.911	**0.686** ± 0.162
w/o Temporal Attention	0.611 ± 0.181	0.798 ± 0.175	3.784 ± 2.999	0.643 ± 0.176
w/o F_{motion}	0.519 ± 0.159	0.764 ± 0.128	5.554 ± 3.158	0.545 ± 0.159
w/o F_{thick}	0.608 ± 0.178	**0.806** ± 0.168	3.883 ± 3.075	0.638 ± 0.171

Accuracy of 3D Infarct Reconstruction. Similar to the Tversky loss in Sect. 2.3, we prioritized the identification of the positive class (i.e., infarct regions), so we emphasized Recall as a key metric. Given that the infarct regions account for only ~ 8.3% of the total LV nodes, we incorporated G Dice score to better evaluate the performance under class imbalance [18]. We evaluated on the test set, and the results are presented in Table 1. CMotion2Infarct-Net achieved

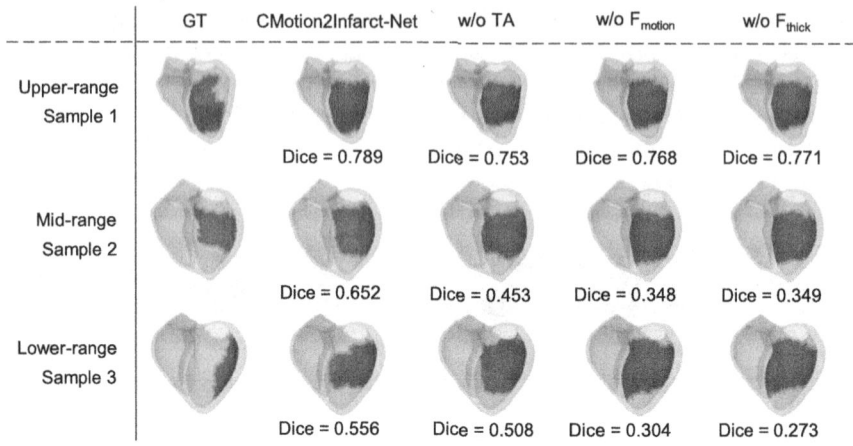

Fig. 4. Comparison of prediction under different Dice scores. TA: Temporal Attention.

a reasonable Recall score of 0.805, indicating strong coverage of scar regions. Although the average Dice score is 0.652, which reflected moderate overlap, the average G Dice score increases to 0.686, highlighting improved performance in identifying small infarct regions across different cases. In addition, ASD suggested that the predicted boundaries are geometrically close to the GT, further validating the spatial precision. Figure 4 illustrates qualitative comparisons between predicted infarct regions and GT across different performance levels. We selected representative test samples from the approximate top, middle, and bottom quartiles based on Dice scores for visualization, with infarct regions highlighted in red. In worse cases, CMotion2Infarct-Net tends to over-predict infarct regions, resulting in some false positives or spatial over-coverage. In contrast, high-performing cases exhibit strong agreement with the ground truth, while median cases show reasonable agreement with slight deviations.

Notably, one potential limitation of supervised approach is that the performance is affected by human expertise. For instance, although most scars should have clearly defined boundaries, some ambiguous regions remain, especially scars near the atria or apex where image quality is worse, which may lead to inter-observer variation. To further evaluate the reliability and stability, we invited an another expert to manually segment 15 cases. A comparison with model predictions is shown in Table 1. The inter-observer Dice score is approximately 0.8, and G Dice score has even reached 0.824. Our goal is to achieve accuracy comparable to human experts. Despite some differences, preliminary results indicate that CMotion2Infarct-Net has achieved reasonable accuracy.

Ablation Study. To evaluate the contributions of key modules and input features to model performance, we designed ablation experiments. The experimental results are shown in Table 1. Regarding input features, we removed the motion feature F_{motion} (which captures myocardial motion) and the thickness feature

F_thick (which reflects myocardial thickness), respectively to evaluate their independent contributions. The experimental results showed that removing either feature led to performance degradation. In particular, when the F_motion was excluded, the all score dropped significantly, indicating that motion was crucial in identifying infarct regions. Although F_thick removal had little impact on Recall, it resulted in lower Dice (G Dice) and higher ASD, suggesting its complementary value. Additionally, we removed the temporal attention module in CMotion2Infarct-Net to assess its effectiveness in modeling temporal dependencies. Both the Dice and Recall scores decreased noticeably, confirming the importance of the Transformer in capturing temporal dynamics. Some of the ablation results are also illustrated in Fig. 4 as a visual comparison with the original predictions. The visual comparisons are consistent with the quantitative results: removing key modules leads to a decline in model performance, particularly in challenging cases (i.e., those around or below the median level).

4 Conclusion

In this work, we proposed an novel framework for automatic 3D infarct geometry reconstruction by combining multi-view cine MRIs. The proposed method fully leverages cardiac motion along the temporal dimension to enable structured representation of infarct regions without contrast agents, offering significant potential for clinical application. The results have showed that the motion information can be accurately extracted from cine MRI and explicitly mapped to the region of infarction. One major challenge is the large heterogeneity in infarct type, morphology, intensity, and spatial distribution. In our future work, We will address these challenges and extend the surface-based infarct representation to a 3D volumetric (tetrahedral) model, enabling its direct use in personalized cardiac simulations and broader digital twin applications for post-infarction patient care.

Acknowledgments. This work was supported by NUS start-up funding to L. Li.

References

1. Alam, M., Wardell, J., Andersson, E., Samad, B.A., Nordlander, R.: Left ventricular regional diastolic dysfunction in patients with first myocardial infarction determined by diastolic motion of the atrioventricular plane. Echocardiography **16**(7, Pt 1), 635–641 (1999)
2. Biffi, C., et al.: 3D high-resolution cardiac segmentation reconstruction from 2D views using conditional variational autoencoders. In: International Symposium on Biomedical Imaging, pp. 1643–1646. IEEE (2019)
3. Cerisano, G., Bolognese, L.: Echo-doppler evaluation of left ventricular diastolic dysfunction during acute myocardial infarction: methodological, clinical and prognostic implications. Ital. Heart. J. **2**(1), 13–20 (2001)

4. Chen, Y., Yang, J., Mercadier, D.S., Le, H., Fua, P.: MedTet: An online motion model for 4D heart reconstruction. arXiv preprint arXiv:2412.02589 (2024)
5. Codreanu, A., et al.: Electroanatomic characterization of post-infarct scars: comparison with 3-dimensional myocardial scar reconstruction based on magnetic resonance imaging. J. Am. Coll. Cardiol. **52**(10), 839–842 (2008)
6. Corral-Acero, J., et al.: The 'Digital Twin' to enable the vision of precision cardiology. Eur. Heart J. **41**(48), 4556–4564 (2020)
7. Dillon, J.R., et al.: An open-source end-to-end pipeline for generating 3D+t biventricular meshes from cardiac magnetic resonance imaging. In: Chabiniok, R., Zou, Q., Hussain, T., Nguyen, H.H., Zaha, V.G., Gusseva, M. (eds.) Functional Imaging and Modeling of the Heart, pp. 372–383. Springer Nature Switzerland, Cham (2025)
8. Feldmann, K.J., Goldstein, J.A., Marinescu, V., Dixon, S.R., Raff, G.L.: Disparate impact of ischemic injury on regional wall dysfunction in acute anterior vs inferior myocardial infarction. Cardiovasc. Revasc. Med. **20**(11), 965–972 (2019)
9. Laumer, F., et al.: Weakly supervised inference of personalized heart meshes based on echocardiography videos. Med. Image Anal. **83**, 102653 (2023)
10. Li, L., et al.: Towards enabling cardiac digital twins of myocardial infarction using deep computational models for inverse inference. IEEE Transactions on Medical Imaging (2024)
11. Li, L., Ding, W., Huang, L., Zhuang, X., Grau, V.: Multi-modality cardiac image computing: a survey. Med. Image Anal. **88**, 102869 (2023)
12. Li, L., et al.: MyoPS: a benchmark of myocardial pathology segmentation combining three-sequence cardiac magnetic resonance images. Med. Image Anal. **87**, 102808 (2023)
13. Liu, J., et al.: Accurate 3D contrast-free myocardial infarction delineation using a 4D dual-stream spatiotemporal feature learning framework. Appl. Soft Comput. **146**, 110694 (2023)
14. Nowosielski, M., et al.: Comparison of wall thickening and ejection fraction by cardiovascular magnetic resonance and echocardiography in acute myocardial infarction. J. Cardiovasc. Magn. Reson. **11**(1), 22 (2009)
15. Polacin, M., et al.: Segmental strain analysis for the detection of chronic ischemic scars in non-contrast cardiac MRI cine images. Sci. Rep. **11**(1), 12376 (2021)
16. Reed, G.W., Rossi, J.E., Cannon, C.P.: Acute myocardial infarction. The Lancet **389**(10065), 197–210 (2017)
17. Smiseth, O.A., Rider, O., Cvijic, M., Valkovič, L., Remme, E.W., Voigt, J.U.: Myocardial strain imaging: theory, current practice, and the future. JACC: Cardiovasc. Imaging **18**(3), 340–381 (2025)
18. Sudre, C.H., Li, W., Vercauteren, T., Ourselin, S., Cardoso, M.J.: Generalised dice overlap as a deep learning loss function for highly unbalanced segmentations. In: Deep Learning in Medical Image Analysis and Multimodal Learning for Clinical Decision Support, pp. 240–248. Springer (2017)
19. Thune, J.J., Solomon, S.D.: Left ventricular diastolic function following myocardial infarction. Curr. Heart Fail. Rep. **3**(4), 170–174 (2006)
20. Ukwatta, E., et al.: Myocardial infarct segmentation from magnetic resonance images for personalized modeling of cardiac electrophysiology. IEEE Trans. Med. Imaging **35**(6), 1408–1419 (2015)
21. Wang, Z.J., et al.: Human biventricular electromechanical simulations on the progression of electrocardiographic and mechanical abnormalities in post-myocardial infarction. EP Eur. **23**(Supplement_1), i143–i152 (2021)

22. Xu, C., Howey, J., Ohorodnyk, P., Roth, M., Zhang, H., Li, S.: Segmentation and quantification of infarction without contrast agents via spatiotemporal generative adversarial learning. Med. Image Anal. **59**, 101568 (2020)
23. Xu, C., et al.: Direct delineation of myocardial infarction without contrast agents using a joint motion feature learning architecture. Med. Image Anal. **50**, 82–94 (2018)
24. Xu, C., et al.: Contrast agent-free synthesis and segmentation of ischemic heart disease images using progressive sequential causal gans. Med. Image Anal. **62**, 101668 (2020)
25. Yang, G., et al.: Contrast-free myocardial scar segmentation in cine MRI using motion and texture fusion. In: 2025 IEEE 22nd International Symposium on Biomedical Imaging (ISBI), pp. 1–5. IEEE (2025)
26. Ye, M., Yang, D., Kanski, M., Axel, L., Metaxas, D.: Neural deformable models for 3D bi-ventricular heart shape reconstruction and modeling from 2D sparse cardiac magnetic resonance imaging. In: Proceedings of the IEEE/CVF International Conference on Computer Vision, pp. 14247–14256 (2023)
27. Yousefi-Banaem, H., Kermani, S., Asiaei, S., Sanei, H.: Prediction of myocardial infarction by assessing regional cardiac wall in CMR images through active mesh modeling. Comput. Biol. Med. **80**, 56–64 (2017)
28. Yuan, X., Liu, C., Wang, Y.: 4D myocardium reconstruction with decoupled motion and shape model. In: Proceedings of the IEEE/CVF International Conference on Computer Vision, pp. 21252–21262 (2023)
29. Zhang, N., et al.: Deep learning for diagnosis of chronic myocardial infarction on nonenhanced cardiac cine MRI. Radiology **291**(3), 606–617 (2019)
30. Zhuang, X.: Multivariate mixture model for myocardial segmentation combining multi-source images. IEEE Trans. Pattern Anal. Mach. Intell. **41**(12), 2933–2946 (2018)

Microvascular Retinal Digital Twins from Non-invasive Clinical Images

Rémi J. Hernandez[1,2](✉) and Wahbi K. El-Bouri[1,2]

[1] Department of Cardiovascular and Metabolic Medicine, University of Liverpool, Liverpool, UK
[2] Liverpool Centre for Cardiovascular Sciences, University of Liverpool and Liverpool Heart and Chest Hospital, Liverpool, UK
remi.hernandez@liverpool.ac.uk

Abstract. Patient-specific haemodynamic simulations have largely focused on the heart chambers or major vessels. This study presents microvascular image-based digital twins of the retinal circulation, including arterial venous and capillary flow in all three macular plexi. The uneven distribution of erythrocytes at vascular junctions was also accounted for in the *in silico* model. Six eyes were modelled, with twenty digital variants each to assess uncertainty in capillary architecture. The model, validated under baseline conditions, also simulates early diabetic retinopathy, revealing increased mean and variability in capillary transit times – potential indicators of retinal hypoxia. Additionally, the model predicts reduced macular perfusion in diabetes, particularly in the central macula, aligning with clinical observations of disease progression.

Keywords: Computational modelling · Retinal haemodynamics · Capillary perfusion · Vascular pathophysiology · Retinal biomarkers

1 Introduction

Visual function induces high metabolic activity in retinal cells, creating consequent oxygen demand. As a result, the retina's oxygen consumption rate is comparable to that of the brain, necessitating a complex, three-laminar network of capillaries developed early after birth. While capillaries remain challenging to visualize and segment, superficial arterioles and venules can be segmented on several image types, such as colour fundus photographs (CFP) and optical coherence tomography angiography (OCTA) [20,30]. Being non-invasive and cheap, these imaging modalities have brought a lot of interest to the study of the retinal vasculature, an interest increased with the advancements in computer vision models, demonstrating the potential of retinal vascular biomarkers. These biomarkers can further be enhanced with simulated haemodynamic parameters, similarly to what was done for example in [22]. In this example, however, only arteries were modelled.

Retinal microcirculation alterations are emerging as biomarkers for retinal and systemic diseases [8,18,20,21]. Understanding how capillary topology affects retinal homeostasis could elucidate the pathophysiology of several retinal disorders [8,21]. However, most mathematical models of the retina adopt a compartmental approach or exclude the capillaries from simulations, thereby overlooking the spatial heterogeneity of the microcirculation [7,14,22,26,28] Additionally, studies often focus on the haemodynamics in the superficial vascular complex (SVC) [22,26,27], despite microvascular changes in the other vascular plexi of the retina – in particular in the macula, the central region of the retina responsible for high acuity colored vision – potentially preceding SVC alterations in conditions like diabetes [8]. While OCTA imaging allows studying these plexi, only few models explicitly include their contribution to haemodynamics [7,11].

The large number of capillaries in the retina make complex 1D or 3D blood flow simulations are often computationally infeasible [2,26]. Most microcirculatory models chose instead to model blood as a steady, single-phased, Newtonian and zero-dimensional flow instead [7,11,15]. This simplification leads to the Poiseuille equation:

$$q = \frac{\pi r^4}{8\eta l} \Delta p, \qquad (1)$$

where q is the volumetric blood flow, r and l the vessel radius and length and Δp the pressure drop between the proximal and distal ends of the vessel. The use of an effective viscosity term $\eta(r, h)$, as a function of vessel radius and discharge haematocrit (h), accounts for the non-Newtonian behavior of blood and the Fåhræus-Lindqvist effect [10,24]. However, the heterogeneous distribution of red blood cells at vascular junctions – also known as plasma skimming – is not accounted for in the standard Poiseuille model and haematocrit is wrongly assumed constant. Empirical models address this, but the computational challenges remain in large networks [12,25]. Accurate haematocrit simulations in realistic networks allows us to faithfully estimate the transit time (TT) probability distribution, a valuable perfusion biomarker [23,32].

To address these limitations, we propose an image-based framework for creating *digital twins* (DT) of the retinal vasculature. Our method constructs digital copies of the superficial temporal vessels and simulates haemodynamics in all three plexi of the macula, incorporating capillaries in the SVC – where they link the digitized arteries and veins–, but also in the intermediate (ICP) and deep capillary plexi (DCP). Using OCTA and CFP images, we created six DTs and a *in-silico* flow model that include plasma skimming and only requires basic blood pressure data as input. The structural and functional model are validated against independent studies. Finally, we present a model of diabetic retinopathy (DR) and show how the early onsets of the disease can affect key haemodynamic parameters of the retina.

2 Materials and Methods

Figure 2 summarizes the steps for creating a DT from an image and simulating haemodynamics in the resulting vascular graph. The following section details the digitization process and haemodynamic model.

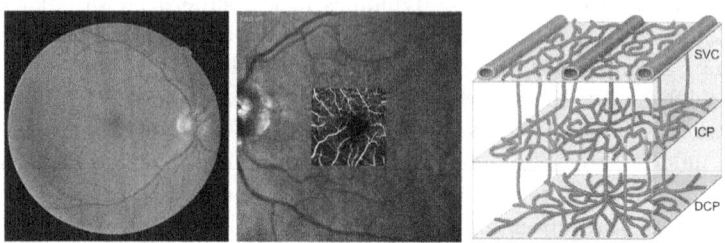

Fig. 1. Left: a colour fundus photograph from the DRIVE dataset, with the macula appearing as a dark, oval area in the centre of the image. Middle: an *en-face* optical coherence tomography (OCT) angiography of the macula, overlaid on an OCT fundus image from an *in-house* dataset. Right: The three-laminar organization of the macular vasculature assumed in this work. Illustration adapted from [13] under CC-BY license.

2.1 Creating Digital Twins of the Vasculature

Data. DTs were created from macular-centered fundus images of patients with no known retinopathy. These images may be OCT images – augmented with an *en-face* OCTA projection of the SVC – or CFP (Fig. 1). These included four CFPs from the publicly available DRIVE dataset [30]. The CFPs are paired with manual, pixel-wise annotations of arteries and veins. Two OCT fundus images from an 'in-house' dataset (approved by NHS REC, No. 22/PR/1189) were included. The OCT imaging modality is less adapted to visualizing the blood vessels and, therefore, fewer branches could be segmented compared to the CFPs. To alleviate this difference, fovea-centred *en-face* OCTA projections of the SVC were overlaid on the fundus image. The registration of the OCT and OCTA images is performed by the Heidelberg Eye Explorer software (version 1.10.12.0; Heidelberg Engineering, Heidelberg, Germany).

Digitization. The vessel annotations of the DRIVE data were automatically converted into graph data, emphasizing on preserving the topology of the vasculature. For the DRIVE data, vessel annotations were automatically converted to graphs using *scikit-image* (version 0.25.0) in Python. First, the pixel-wise annotations were converted to centreline annotations using the SKELETONIZE function. Vessel diameters were obtained using the MEDIAL_AXIS function, used only for the diameter map due to artefacts in the skeletonizaton with this function. Vessel centrelines were then converted to graphs using the PIXEL_GRAPH function, adding vertices for pixels and edges for adjacent ($\|x_u - x_v\|_\infty \leq 1$)

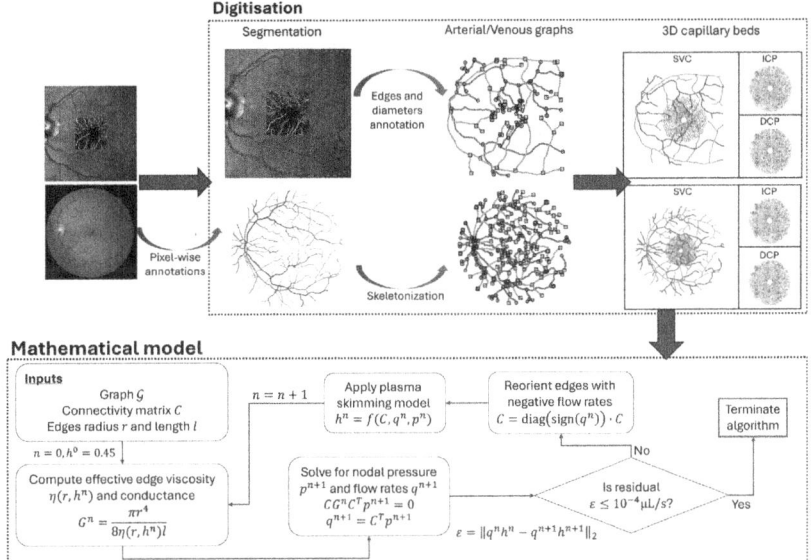

Fig. 2. Summary of the methodology to create digital twins and flowchart of the algorithm used for the blood flow model.

annotated pixels. Edge diameters were averaged from connected vertices' diameters.

For the OCT/OCTA data, graphs of arteries/veins centreline were manually created using an *in-house* Python tool. Diameters were measured for all vessels forming vascular junctions and kept constant between two junctions.

At this stage, the segmented arterial and venous circulation form two disconnected graphs. The largest artery and the largest vein within the optic disc region were selected as the inlet (central retinal artery, CRA) and outlet (central retinal vein, CRV), respectively, of each graph. These inlets and outlets correspond to the central retinal vessels and act as the roots of the arterial and venous graphs – therefore being the only point of contact with the systemic circulation.

Interplexi Connections. This study models the macular vasculature as a three-layered structure (Fig. 1), with arterioles and venules present only in the SVC. Connections to deeper plexi are based on *post-mortem* data [1], which reported ≈ 11 arterial and venous connections per mm^2 between the SVC and the ICP-DCP. To replicate this, up to eleven randomly selected arteriolar and venular segments were randomly selected within each $1\,mm^2$ region of the DT's macula. Each segment bifurcates at its midpoint toward the DCP (depth $z_{DCP} = 180\,\mu m$) with an intermediate vertex (depth $z_{ICP} = 135\,\mu m$) connecting to the ICP.

Capillary Beds. Capillary networks of the SVC, ICP and DCP are generated using 2D Voronoi tessellations, seeded by N_{seeds} points within a 5 mm macular disk, excluding a 250 μm radius around the center of the fovea. Each seed is enclosed by a polygon whose edges are added to the vascular graph as capillaries. In the SVC, connections are made where capillaries intersect arterioles or venules, and end points of the segmented arterioles and venules are linked to their nearest capillary. Since the tessellation tiles space into polygons, interplexi vessels can only cross one polygon, of which up to four of vertices are randomly connected to the interplexi vessel [31]. Capillary diameters d follow a normal distribution (mean 8 μm, SD 0.1 μm) [5,6].

2.2 Haemodynamics Model

The model assumes Poiseuille flow (Equation (1)), with effective viscosity $\eta = \eta_{pl}\eta_{eff}(r,h)$, where $\eta_{pl} = 1.2$ cP is the viscosity of plasma. Let $\mathcal{G} = (V, E)$ be the vascular graph with vertex V and edge set $E \subset V \times V$ and let C be its incidence matrix. Let $\mathbf{G} = \text{diag}\left(\{\pi r_e^4/(8\eta_e l_e)\}_{e \in E}\right)$ be the diagonal conductance matrix. Mass conservation at internal vertices yields for pressure p:

$$\mathbf{CGC}^T\mathbf{p} = 0. \tag{2}$$

The macular network is fully connected and has no terminal vertices. In the periphery, unconnected segmented arterioles and venules (with in/out-degree 0) act as sources and sinks, requiring additional constraints. As the retina is an end-organ, conservation of mass between the CRA and CRV is enforced via $q_{CRA} = q_{CRV}$, modifying the row corresponding to the CRV vertex in Equation (2). A pressure $p_{CRA} = \frac{2}{3}MAP$ is set at the CRA, based on a mean arterial pressure (MAP) of 94 mmHg. Terminal venule pressures are assigned using a diameter-pressure relationship from a symmetric branching tree [31].

Terminal arterioles are implicitly extended with a symmetric branching tree (up to 8 generations) through the following boundary condition:

$$p_v = p_{cap} + \sum_{e \to (\cdot,v)} Z_v \frac{\Delta p_e}{R_e}, \tag{3}$$

where p_v is the unknown terminal arteriole pressure, $p_{cap} = 22.5$ mmHg, Z_v is the impedance of the appended tree, and the sum is over incoming edges [31].

Plasma Skimming Model. The model included a phenomenological model of plasma skimming [25]. Introducing the fraction of whole blood (Q_B) and erythrocytes (Q_E) flow between feeding and daughter branches:

$$FQ_B^i = \frac{Q_B^i}{Q_B^f} \text{ and } FQ_E^i = \frac{Q_E^i}{Q_E^f}, \text{ with } i \in \{a,b\}, \tag{4}$$

the phase separation law is the following parameterized relation:

$$\text{logit}\left(FQ_E^i\right) = A + B\,\text{logit}\left(\frac{FQ_B^i - X_0}{1 - 2X_0}\right), \tag{5}$$

if $X_0 < FQ_B^i < 1 - X_0$ else $FQ_E^i = 0$ (if $FQ_B^i \leq X_0$) or $FQ_E^i = 1$ (if $FQ_B^i \geq 1 - X_0$), where $\text{logit}(x) = \ln[x/(1-x)]$. Here X_0, A and B are non-dimensional variables dependent on the feeding vessel diameter and haematocrit. The three length parameters $\underline{A}$, $\underline{B}$ and $\underline{X_0}$ are empirical values scaled from rat to human vasculature and can be found in [19]. The above model describes the phase separation at bifurcations only. In other junctions, the split of erythrocytes is assumed even, namely, $FQ_E^i = 1/k$ for a junction with k distal branches. Values of haematocrit are clipped between 0.2 and 0.8 after each iteration [19].

Implementation. Equation (2) can be adapted to include the boundary conditions using the $n_v \times n_v$ diagonal decision matrix $\mathbf{D} = \text{diag}(d_1, \ldots, d_n)$ where $d_i = 1$ if vertex i has in-degree 0, else $d_i = 0$ and n_v is the number of vertices in the vascular graph. The full system then reads:

$$(I - D) \cdot \mathbf{C} \cdot \text{diag}(\mathbf{G}) \cdot \mathbf{C}^T \mathbf{p} + \mathbf{K}p = \mathbf{S}, \tag{6}$$

where I is the identity matrix of size n_v and $\mathbf{K} \in \mathbb{R}^{n_v \times n_v}$ and $\mathbf{S} \in \mathbb{R}^{n_v}$ implement the boundary conditions Equation (3).

With plasma skimming, haematocrit becomes a variable, introducing non-linearity. Varying haematocrit distributions can also induce flow reversal in some vessels. An equilibrium of the pressure-haematocrit state is found using a Picard fixed-point iteration, accommodating for flow direction change. Convergence is reached when the flow difference falls below $\epsilon = 10^{-4}\,\mu\text{L}\,\text{s}^{-1}$. The algorithm is outlined in Fig. 2.

The transit time distribution (TTD) was estimated via a flow-biased random walk from the CRA to the CRV. Vessels with erythrocyte flow $< 10^{-8}\,\mu\text{L}\,\text{s}^{-1}$ were excluded due to negligible erythrocyte traversal probability. The TTD was characterized by its mean (MTT) and standard deviation (capillary transit time heterogeneity, CTH).

3 Results and Discussion

3.1 Validation

Six eyes were digitized: four from CFP images (namely, *22_training*, *24_training*, *26_training*, *35_training*) and two from OCT/OCTA images (namely, *A010* and *A148*). Capillaries, not reliably segmentable from these images, were generated. Capillary density, estimable from OCTA data, varies with image processing, and capillaries are invisible in CFPs. To address this uncertainty, 20 DT versions per individual were generated, each with N_{seeds} sampled from normal distributions with mean±std matching empirical measurements: 6000±1000 (SVC), 3000±500

(ICP) and 2500 ± 500 (DCP). The i-th model maintains consistent values of N_{seeds} across individuals.

Mean capillary density in the SVC, ICP and DCP were 0.33 ± 0.03, 0.25 ± 0.01 and 0.22 ± 0.02, respectively, aligning with OCTA and histology values (0.30 to 0.50, 0.17 to 0.25, 0.16 to 0.26 [4,17]). Similarly, inter-capillary distances – $(37.3\pm3.1), (45.4\pm1.7), (47.4\pm1.8)\,\mu$m in the SVC, ICP and DCP, respectively – were close to values from [6] (47, 54 and 52 μm).

Total retinal blood flow (TRBF) (i.e., q_{CRA}) was (43.28 ± 12.19) μL min^{-1}, similar to what has been measured *in-vivo* in the retina (30 μL/min to 50 μL/min [9]). The shortest transit times (0.3 s to 2.5 s) were comparable with experimental measurements of arterio-venous passage (namely, (1.46 ± 0.57) s [3]). Additionally, MTT ranged from 4.8 s to 9.9 s, which also aligns with measured MTT $((4.9\pm1.9)$ s), though the value may vary by a few seconds depending on the technique used [3].

3.2 A Model of Diabetic Retinopathy

Diabetes is a disease affecting both the macrocirculation and microcirculation. In the retina, capillary rarefaction constitutes one of the sign of DR. To simulate the advancement of diabetic symptoms, random capillaries are removed from the macula and the structured trees at arteriolar outlets, while maintaining the global topology in the graph. Namely, a capillary is dropped if removing it does not create new terminal vessels for the graph. The stages of DR are labeled *intermediate* and *advanced* based on the percentage of capillary loss, ranging for the different plexi from 1.29 % to 2.13 % in the intermediate stage and from 3.04 % to 4.87 % in the advanced case. At the arteriolar outlets, the capillary drop rates were 2 and 5 %.

Effects on Global Haemodynamics. Diabetes increased TRBF as capillaries were removed, reducing overall vascular resistance in macular capillary beds. However, fixed pressure boundary conditions should also be adjusted to reflect disease mechanisms for accurate diseased TRBF estimation. It should be noted that the global decrease in vascular resistance induces increased TRBF but may be accompanied with localized increase in capillary resistance. The eyes created from OCTA data showed several times higher increases in comparison to the eyes created from CFP data.

Figure 3 shows a shift in TTD with disease progression. MTT (40 %) and CTH (120 %) rose in the advanced stages, indicating higher oxygen perfusion in some regions (slower TT) but faster TT and reduced oxygen extraction in others. Several studies suggested that higher blood flow may not ensure adequate oxygen perfusion and could coincide with reduced tissue oxygen tension with capillary disruption [23,32]. Higher TRBF may reflect compensatory hyperemia in mild capillary dysfunction with increased capillary CTH.

Figure 4 shows the pressure change in the advanced DR stage. Reduced capillary density appears to locally increase flow resistance, leading to rapid

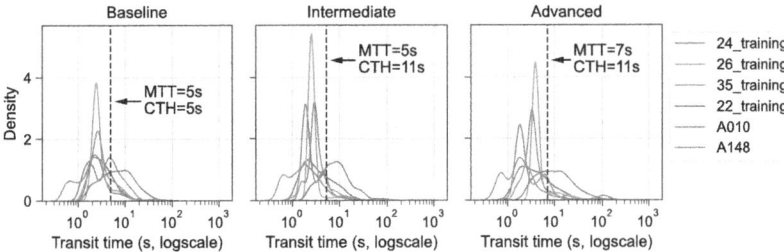

Fig. 3. Plot of the probability density functions of transit times with the advancement of diabetic symptoms. Each line corresponds to a single digital twin. Vertical dashed lines show the means of MTT and CTH among the six individuals plotted.

dissipation of the high pressure in arterioles and resulting in a low, uniform macular capillary pressure. Although average pressure decreased by 20% in DR, the proportion of capillaries with elevated pressure (up to 45%) increased from 5% (intermediate) to 8% (advanced). This localized pressure rise, along with increased capillary permeability, may explain the localized manifestation of macular edema [16]. Small reductions in capillary density significantly lowered perfusion pressure in the nasal macula, especially near nasal vessels, unlike those from the superior or inferior quadrants. Erythrocyte flow declined away from arterioles and venules, creating hypoperfused zones, notably around the foveal avascular zone (FAZ), which is primarily capillary-supplied. These findings suggest that increased resistance and lower capillary pressure shunt flow toward draining venules, contributing to perifoveal flow increases and supporting the hypothesis that FAZ enlargement is driven by ischemia [29].

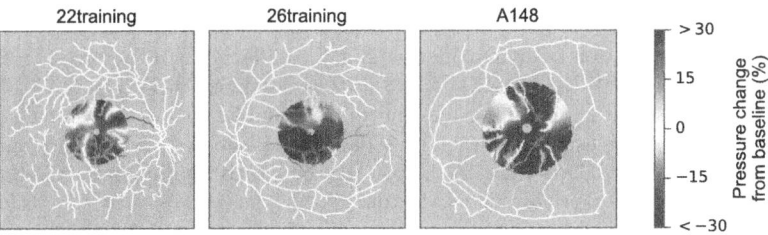

Fig. 4. Plot of the pressure change in the SVC between the normal conditions and the advanced diabetic state for a subset of the digital twins.

4 Conclusion

Despite the ability to visualize the retinal vasculature at the smallest scale, the associations between morphology and haemodynamic functions remains elusive in age and disease. We proposed a new framework for creating DTs of the retinal microcirculation from non-invasive images. The structure and function of the DTs closely reproduced the morphometry and haemodynamics of the retina. We proposed a simple model of diabetic retinopathy which provides new insights into the capillary haemodynamic, such as changes in TTD.

The proposed DT pipeline can serve as a useful tool to study TTD in realistic networks and prove its value as a biomarker of retinal perfusion. Furthermore, the framework can enhance classical biomarkers, improving early detection, enabling risk stratification and informing closer monitoring of patients in disease such as diabetes.

References

1. An, D., et al.: Three dimensional characterization of the normal human parafoveal microvasculature using structural criteria and highresolution confocal microscopy. Invest. Ophthalmol. Vis. Sci. **61**(10), 3 (2020). https://doi.org/10.1167/iovs.61.10.3
2. Bernabeu, M.O., et al.: Computer simulations reveal complex distribution of haemodynamic forces in a mouse retina model of angiogenesis. J. R. Soc. Interface **11**(99) (2014). https://doi.org/10.1098/rsif.2014.0543
3. Bjärnhall, G., et al.: Analysis of mean retinal transit time from fluorescein angiography in human eyes: normal values and reproducibility. Acta Ophthalmol. Scand. **80**(6), 652–655 (2002). https://doi.org/10.1034/j.1600-0420.2002.800617.x
4. Chan, G., et al.: Quantitative morphometry of perifoveal capillary networks in the human retina. Invest. Ophthalmol. Vis. Sci. **53**(9), 5502–5514 (2012). https://doi.org/10.1167/iovs.12-10265
5. Chan, G., et al.: In vivo optical imaging of human retinal capillary networks using speckle variance optical coherence tomography with quantitative clinicohistological correlation. Microvasc. Res. **100**, 32–39 (2015). https://doi.org/10.1016/j.mvr.2015.04.006
6. Chandrasekera, E., et al.: Three-dimensional microscopy demonstrates series and parallel organization of human peripapillary capillary plexuses. Invest. Ophthalmol. Vis. Sci. **59**(11), 4327–4344 (2018). https://doi.org/10.1167/iovs.18-24105
7. Chiaravalli, G., et al.: A multiscale/multiphysics model for the theoretical study of the vascular configuration of retinal capillary plexuses based on OCTA data. Math. Med. Biol. **39**(1), 77–104 (2022). https://doi.org/10.1093/imammb/dqab018
8. Chua, J., et al.: Optical coherence tomography angiography of the retina and choroid in systemic diseases. Prog. Retin. Eye Res. **103**, 101292 (2024). https://doi.org/10.1016/j.preteyeres.2024.101292
9. Doblhoff-Dier, V., et al.: Measurement of the total retinal blood flow using dual beam Fourier domain Doppler optical coherence tomography with orthogonal detection planes. Biomed. Opt. Express **5**(2), 630–642 (2014). https://doi.org/10.1364/BOE.5.000630, publisher: Optica Publishing Group
10. Fåhræus, R., Lindqvist, T.: The viscosity of the blood in narrow capillary tubes. Am. J. Physiol. **96**(3), 562–568 (1931). https://doi.org/10.1152/ajplegacy.1931.96.3.562
11. Fry, B.C., et al.: Predicting retinal tissue oxygenation using an image-based theoretical model. Math. Biosci. **305**, 19 (2018). https://doi.org/10.1016/j.mbs.2018.08.005

12. Gould, I.G., Linninger, A.A.: Hematocrit distribution and tissue oxygenation in large microcirculatory networks. Microcirculation **22**(1), 1–18 (2015). https://doi.org/10.1111/micc.12156
13. Haddad, C., et al.: An OCT-A analysis of the importance of intermediate capillary plexus in diabetic retinopathy: a brief review. J. Clin. Med. **13**(9), 2516 (2024). https://doi.org/10.3390/jcm13092516
14. Hernandez, R.J., et al.: Advancing treatment of retinal disease through in silico trials. Prog. Biomed. Eng. **5**(2), 022002 (2023). https://doi.org/10.1088/2516-1091/acc8a9
15. Hernandez, R.J., et al.: Linking vascular structure and function: image-based virtual populations of the retina. Invest. Ophthalmol. Visual Sci. **65**(4), 4040 (2024). https://doi.org/10.1167/iovs.65.4.40
16. Klaassen, I., et al.: Molecular basis of the inner blood-retinal barrier and its breakdown in diabetic macular edema and other pathological conditions. Prog. Retin. Eye Res. **34**, 19–48 (2013). https://doi.org/10.1016/j.preteyeres.2013.02.001
17. Lavia, C., et al.: Retinal capillary plexus pattern and density from fovea to periphery measured in healthy eyes with swept-source optical coherence tomography angiography. Sci. Rep. **10**(1) (2020). https://doi.org/10.1038/s41598-020-58359-y
18. López-Cuenca, I., et al.: Retinal vascular study using OCTA in subjects at high genetic risk of developing Alzheimer's disease and cardiovascular risk factors. J. Clin. Med. **11**(11), 3248 (2022). https://doi.org/10.3390/jcm11113248
19. Lorthois, S., et al.: Simulation study of brain blood flow regulation by intra-cortical arterioles in an anatomically accurate large human vascular network: Part i: Methodology and baseline flow. Neuroimage **54**(2), 1031–1042 (2011). https://doi.org/10.1016/j.neuroimage.2010.09.032
20. Ma, Y., et al.: ROSE: a retinal OCT-Angiography vessel segmentation dataset and new model. IEEE Trans. Med. Imaging **40**(3), 928–939 (2021). https://doi.org/10.1109/tmi.2020.3042802
21. Narnaware, S.H., et al.: Vessel density changes in choroid, chorio-capillaries, deep and superficial retinal plexues on OCTA in normal ageing and various stages of age-related macular degeneration. Int. Ophthalmol. **43**(10), 3523–3532 (2023). https://doi.org/10.1007/s10792-023-02758-3
22. Orlando, J.I., et al.: Towards a Glaucoma Risk Index Based on Simulated Hemodynamics from Fundus Images, pp. 65–73. Springer International Publishing (2018). https://doi.org/10.1007/978-3-030-00934-2_8
23. Payne, S.J., et al.: Transit time mean and variance are markers of vascular network structure, wall shear stress distribution and oxygen extraction fraction. Biomech. Model. Mechanobiol. (2025). https://doi.org/10.1007/s10237-025-01959-2
24. Pries, A.R., et al.: Resistance to blood flow in microvessels in vivo. Circ. Res. **75**(5), 904–915 (1994). https://doi.org/10.1161/01.res.75.5.904
25. Pries, A.R., et al.: Structural response of microcirculatory networks to changes in demand: information transfer by shear stress. Am. J. Physiol. Heart Circ. Physiol. **284**(6), H2204–H2212 (2003). https://doi.org/10.1152/ajpheart.00757.2002
26. Rebhan, J., et al.: A computational framework to investigate retinal haemodynamics and tissue stress. Biomech. Model. Mechanobiol. **18**(6), 1745–1757 (2019). https://doi.org/10.1007/s1023701901172y
27. Sala, L., et al.: Multiscale modeling of ocular physiology. JMO **2**(1), 1218 (2018). https://doi.org/10.35119/maio.v2i1.52

28. Sala, L., et al.: Ocular mathematical virtual simulator: a hemodynamical and biomechanical study towards clinical applications. J. Coupled Syst. Multiscale Dyn. **6**(3), 241–247 (2018). https://doi.org/10.1166/jcsmd.2018.1165
29. Sim, D.A., et al.: Patterns of peripheral retinal and central macula ischemia in diabetic retinopathy as evaluated by ultra-widefield fluorescein angiography. Am. J. Ophthalmol. **158**(1), 144-153.e1 (2014). https://doi.org/10.1016/j.ajo.2014.03.009
30. Staal, J., et al.: Ridge-based vessel segmentation in color images of the retina. IEEE Trans. Med. Imaging **23**(4), 501–509 (2004). https://doi.org/10.1109/tmi.2004.825627
31. Takahashi, T., et al.: A mathematical model for the distribution of hemodynamic parameters in the human retinal microvascular network. J. Biorheol. **23**(2), 77–86 (2009). https://doi.org/10.1007/s12573-009-0012-1
32. Østergaard, L.: Blood flow, capillary transit times, and tissue oxygenation: the centennial of capillary recruitment. J. Appl. Physiol. **129**(6), 1413–1421 (2020). https://doi.org/10.1152/japplphysiol.00537.2020

Validating Digital Twins with Tactile-Visual Liver Phantoms for Robot-Assisted Surgical Workflows

Chengzheng Mao, Ying Zhen Tan, and Yujia Gao(✉)

Academic Informatics Office, National University Health System (NUHS),
Singapore, Singapore
`yujiagao@nus.edu.sg`

Abstract. This study presents a novel physical liver phantom that replicates the mechanical and visual properties of healthy, fatty, and fibrotic liver tissues to support verification, validation, and uncertainty quantification in digital twin frameworks for liver surgery. Simulating these three common liver conditions improves phantom fidelity for surgical training and model validation. We evaluated multiple materials (agar, gelatin, and konnyaku) for phantom fabrication and conducted a twelve-question survey with 25 surgeons to optimize formulations based on tactile realism and visual appearance. Standard surgical interactions (compression, pinching, pulling and cutting) were performed on both phantoms and fresh porcine liver models using the da Vinci® Xi surgical system. Stereo endoscopic video recordings enabled a comparative analysis of visual fidelity and tissue deformation under surgical manipulation. Mechanical properties (compressive and tensile Young's modulus) were quantitatively measured under mechanical testing and compared with real tissue measurements in literature. The resulting phantom, validated through surgeon feedback and mechanical testing, provides a tangible platform for physical model validation and surgeon-in-the-loop simulations in digital twin applications. This work advances digital twin methodologies by offering a realistic testbed for model calibration and verification, bridging computational models and real surgical scenarios.

Keywords: digital twin · liver phantom · robotic assisted · surgical simulation

1 Introduction

To advance surgical assistance in liver surgery, we propose integrating physical liver phantoms with digital twin technology. While personalized digital twins can simulate patient-specific liver physiology, the challenge lies in acquiring comprehensive intra-operative data from human livers needed for highly realistic models. To

---

Supplementary Information The online version contains supplementary material available at https://doi.org/10.1007/978-3-032-07694-6_3.

address this, we aim to systematically use physical models including ex-vivo tissues and gel models to calibrate, verify and refine our digital twin simulations. By combining quantitative assessments with traditional surgical feedback, and by referencing comparative analyses with existing material research, our approach strengthens the connection between physical and virtual modeling, ultimately supporting the creation of clinically relevant and robust digital liver twins of the human liver. Several studies have demonstrated the efficacy of using liver phantoms in procedural navigation and imaging validation. The "Virtual Liver" digital twin enables drug discovery applications [24], while "iPhantom" framework automates patient-specific phantom creation for individualized dosimetry assessments in CT imaging [8]. These systems show that digital twins provide real-time 3D reconstruction and enables dynamic modeling of the tissue for surgical guidance [6]. Further integration with augmented reality demonstrates a 2.22.8 mm accuracy for tumor puncture [22] and precise tumor/vessel localization in laparoscopic resection [1].

Despite these advances, soft tissue deformation still presents critical challenges for image-guided surgery due to difficult localization of deformation changes [10], though marker-based methods effectively handle larger deformations [9]. Human-to-phantom frameworks improves registration robustness through realistic deformation simulations for image guided liver surgery [5], while dynamic models such as digital twins to address intraoperative accuracies that arise from soft tissue deformations [20,21]. Simulating the real-time behaviour of deformable elastic bodies remains a challenging and actively explored field. Computational modeling relies predominantly on FEM modelling for accurate organ deformation [13,14,16,17], with optimizations enabling real-time applications [2,3,27]. However, significant computational constraints are required to run FEM in real time and difficulties in accommodating unclear or variable material properties and boundary conditions poses further challenges [15,27]. AI-driven systems have enhanced real-time deformation tracking [12] and respiratory motion measurement [23]. This suggests that a convergence of phantoms, deformation modeling, and digital twins could address surgical complexities, facilitating precise registration and improved patient outcomes.

Due to the inherent challenges in directly obtaining accurate material properties and boundary conditions from human liver tissue, alternative approaches are essential for the Verification, Validation, and Uncertainty Quantification (VVUQ) of liver digital twins. The use of physical liver phantoms offers a controlled and reproducible platform for investigating anatomical structures and physiological behaviors analogous to those of the actual human organ. By monitoring these phantoms under experimental conditions, essential data can be obtained to validate and calibrate the digital twin models. This approach facilitates the development of digital representations that more faithfully reproduce the physical and functional characteristics of the human liver. Hence, to address these challenges and advance digital twin validation, we developed novel liver phantoms using accessible materials with tunable mechanical properties.

2 Material and Methods

2.1 Liver Phantom Material Preparation

Based on literature review [4,7,18,19,25], agar, gelatin, and konnyaku were selected as base materials for liver phantom fabrication due to their tunable mechanical properties and widespread use in soft tissue simulation. Various gel formulations combining these materials with water and sugar were developed to replicate liver tissue characteristics (Table 1). The mixtures were continuously stirred for 5 min each with and without heat. After cooling the mixtures to 4050°C, they were poured into 50 mm diameter plastic cups. Parallel samples were prepared using cups pre-treated with food-grade oil or WD-40 to evaluate demolding effectiveness. Samples were covered with cling film, partially solidified at room temperature, then refrigerated to let it set.

Table 1. Candidate formulations for liver phantom fabrication

	Sample No.	Ingredient (g)	Sugar (g)	Water (g)	Observation	
					WD-40	Oil
Agar	A1	3.0	75	300	Rough surface (15mm dent) from air bubbles	Rough surface (2mm dent) from air bubbles
	A2	3.0	0	300	Smooth surface	Rough surface (1mm dent) from air bubbles
	A3	6.0	0	300	Rough surface (25mm dent) from air bubbles	Rough surface (2mm dent) from air bubbles
	A4	3.0	0	600	Smooth surface	Smooth surface
	A5	2.0	0	300	Smooth surface	Smooth surface
	A6	3.0	25	300	Rough surface (1mm dent) from air bubbles	Smooth surface
Gelatin	G1	9.0	0	300	Smooth surface; hardly demoldable	Smooth surface; hardly demoldable
	G2	4.5	0	300	Smooth surface; barely demoldable despite multiple attempts	Smooth surface; barely demoldable despite multiple attempts
	G3	3.0	0	300	Smooth surface; reluctantly came loose but remained partially stuck	Smooth surface; reluctantly came loose but remained partially stuck
Konnyaku	K1	3.0	0	300	Smooth surface; easy to demold; water separate out after 2 days	Smooth surface; easy to demold; water separate out after 2 days
	K2	1.5	0	300	Smooth surface; easy to demold; water separate out after 2 days	Smooth surface; easy to demold; water separate out after 2 days

2.2 Material Selection User Study

To determine the optimal mixture for creating the liver phantom, an initial study was conducted with 3 expert liver surgeons to identify formulations most similar to healthy, fatty, or fibrotic liver states (Table 1). Six shortlisted mixtures were prepared in standardized plastic cups under identical conditions, cooled, and refrigerated for solidification. A subsequent user study of 10–12 questions was conducted with these mixtures with surgeons and residents under the Hepatobiliary and General Surgery department. Participants evaluated mixture samples through tactile assessment in a blinded study design without prior knowledge of formulations. Survey responses determined the most suitable mixture for each liver state by synthesizing feedback on samples that most accurately mimicked varied liver conditions. Results informed fabrication of full liver phantom volumes for subsequent validation testing against actual liver tissue.

2.3 Mechanical Testing

To characterize the mechanical fidelity of liver phantom material properties, compression and tensile tests were conducted using an Instron® universal testing system (Model 5500) (Fig. 1). Test specimens prepared with the K2 formulation, selected from previous user studies to replicate healthy liver tissue deformation behavior for surgical simulation and digital twin applications. For compression testing, cylindrical specimens (50 mm diameter, 50 mm or 25 mm height) were cast in PLA molds 3D-printed using a Creality® Ender 3 S1 Plus printer. Two specimens were prepared per height category. Compression plates and load cell were covered with cling film to prevent moisture damage. For tensile testing, dog-bone specimens were cast using PLA+ molds. Gauge sections measured 40 mm and 50 mm length with 5 mm and 4 mm thickness, respectively. Tensile fixtures and load cell were covered with cling film, with tissue paper around grip regions preventing slipping. All samples were tested to failure for comprehensive stress-strain data to calculate Young's and tensile modulus. This characterization ensures accurate liver phantom modeling in digital twin environments and robot-assisted surgical workflows.

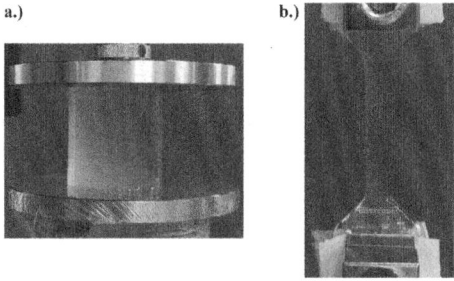

Fig. 1. Mechanical testing setup. a.) Compression test; b.) Tensile test.

2.4 Robotic Assisted Surgery Simulation and Verification

The da Vinci® Xi Surgical System (Model PS4000) was used to assess liver phantom deformation behavior and digital twin validation feasibility through comparative experiments. A 3D liver mesh model reconstructed from donor CT scans using TotalSegmentator [26] was imported into Autodesk Fusion 360 (version 2602.0.71) to create a negative mold. The PLA-printed mold was sealed with epoxy and duct tape before gel casting and refrigerated overnight using the optimal healthy liver formulation. Fresh porcine liver served as a biological reference, refrigerated immediately to preserve mechanical properties. Both specimens underwent four manipulative actions: cutting, compressing, pulling, and pinching (grasping without translation). Real-time video capture documented mechanical responses under robotic manipulation, enabling qualitative and semi-quantitative analysis of deformation behavior, texture response, and structural integrity. Finite Element Analysis (FEA) using Abaqus/CAE 2023 simulated the deformation patterns for comparison with the results from both gel phantom and porcine liver specimens, validating the digital twin development.

2.5 Ultrasound Study

To further validate the phantom's internal structure and its ability to detect tumors, a smaller liver mold was 3D printed and filled with the same mixture representing healthy liver tissue. Additionally, spherical tumor phantoms were created using 3D printed molds. These tumor spheres, made from a material mixture similar to fibrotic liver tissue, were prepared several days prior to creating the main liver phantom. Food coloring was added to the liver phantom mixture, which was then cooled to approximately 45°C before being poured into the mold. The tumor phantom had been pre-positioned at the bottom of the liver mold, where it was designed to float upward due to buoyancy as the liquid phantom material was poured into the mold.

3 Results and Discussion

3.1 Liver Phantom Material Feedback and User Study Feedback

Based on expert consensus (Table 2 and Fig. 2), 6 formulations were shortlisted for clinician evaluation: No.1 (A4), No.2 (A5), No.3 (K2), No.4 (G2), No.5 (K1) and No.5 (K2 with sugar). Survey results revealed that the clinicians felt K2 most closely resembled healthy liver tissue, exhibiting optimal mechanical compliance, structural stability, moderate deformation resistance, and a smooth, moist surface similar to in vivo properties. A5 mimicked fibrotic liver's firmer texture while K1 replicated fatty liver's softer consistency, suggesting these formulations could potentially be used to simulate pathological states for enhanced digital twin fidelity. Sugar inclusion showed no significant differences in tactile feedback or mechanical properties versus sugar-free variants but increased stickiness and reduced shelf life due to microbial susceptibility. Demolding assessment revealed WD-40 enabled superior release with fewer surface defects but produced residual odor and stickiness unsuitable for clinical settings. Food-grade oil provided a cleaner alternative despite lower efficiency, suggesting application-specific usage: WD-40 for prototyping, food-grade oil for clinical training.

Table 2. Texture Evaluation Results for Gel Samples by Hepatobiliary Experts

Professionals	Comments
Head & Senior Consultant	Healthy Liver - K2 & G2; Fatty Liver - K1; Fibrotic Liver - A4 & A5.
Senior Consultant	Healthy Liver - K2; Fatty Liver - K1 but softer; Fibrotic Liver - A5 & A6
Consultant	Healthy Liver - K2 & A4 (if less bouncy); Fatty liver - K1; Fibrotic Liver - A5 & A6

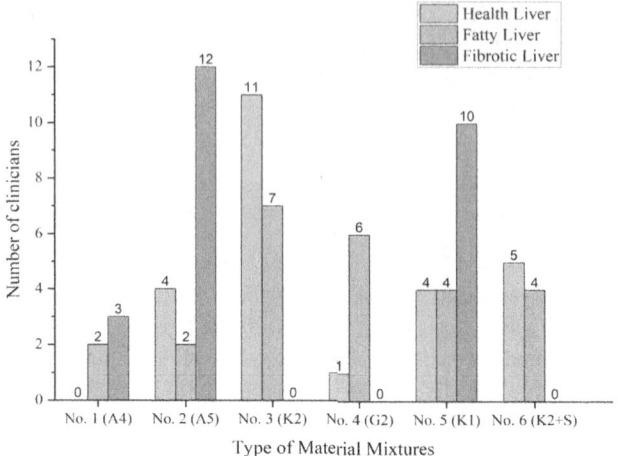

Fig. 2. Clinician voting results (n = 25) for phantom formulations best representing healthy, fatty, and fibrotic liver tissue based on tactile similarity assessment.

3.2 Quasi-Static Compression and Tensile Test

Based on the quasi-static compression and tensile test results, which are summarized in Tables 3 and 4, the corresponding stressstrain curves for both test modes are presented in Fig. 3. All tested specimens exhibited consistent trends within the expected range of soft tissue-like behavior. A linear regression was applied to the linear elastic region of each curve to determine the material's elastic moduli. The resulting Pearson correlation coefficients (r) and coefficients of determination (R^2) were both close to 1, indicating excellent linearity and confirming the appropriateness of the linear fit.

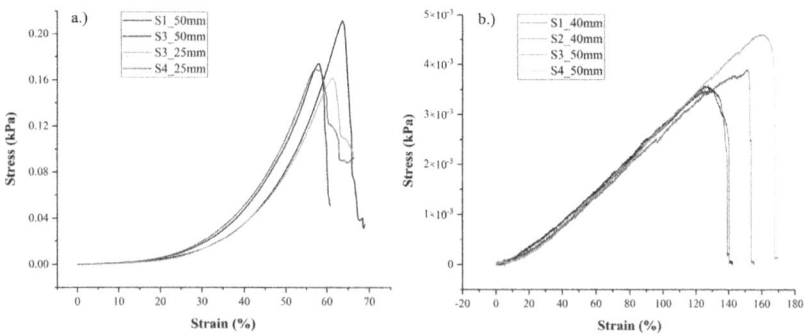

Fig. 3. Stress-strain curve for mechanical tests. a.) Compression test; b.) Tensile test

The liver phantom material demonstrated a compressive modulus of approximately 0.761 kPa and a tensile modulus of approximately 3.297×10^{-3} kPa. The compressive modulus aligns closely with published values for healthy human liver tissue [11], validating the suitability of the K2 formulation for biomechanical liver simulation. These mechanical parameters serve as baseline reference values for material optimization and input essential for developing our digital twin and to support integration into surgical navigation workflows. This mechanical fidelity is critical for achieving tactile realism and accurate visual deformation within digital twin environments. The ability to fabricate consistent, reproducible, and anatomically informed liver phantoms provides a practical platform for verification and uncertainty quantification in digital twin development. The combination of 3D liver modeling, mold fabrication, and controlled gel casting forms a viable pathway for producing physical ground truths to virtual organs, crucial for digital twin applications in personalized medicine and surgical planning.

Table 3. Linear fit for compressive Young's modulus from compression test

Sample	Diameter (mm)	Height (mm)	$E_{compression}$ (kPa)	Pearson's r	R-Square
No. 1	50	50	0.744	0.99263	0.98532
No. 2	50	50	0.818	0.98857	0.97726
No. 3	50	25	0.697	0.99575	0.99152
No. 4	50	25	0.786	0.99722	0.99444
	Average		**0.761**		

Table 4. Linear fit for tensile Young's modulus from tension test

Sample	Width (mm)	Length (mm)	Thickness (mm)	$E_{tension}$ ($\times 10^{-3}$ kPa)	Pearson's r	R-Square
No. 1	10	40	5	3.314	0.99957	0.99915
No. 2	10	40	5	3.165	0.99948	0.99896
No. 3	10	50	4	3.362	0.99796	0.99593
No. 4	10	50	4	3.346	0.99947	0.99893
		Average		**3.297**		

3.3 Robotic Assisted Surgery Simulation and Verification

The comparative assessment using the surgical system validated the liver phantom's realism. K2 gel phantoms demonstrated deformation responses to cutting, pinching, pulling, and compression broadly similar to fresh porcine liver tissue. While responses were not entirely identical, the phantom's surface response, structural compliance, and visual feedback during robotic manipulation were deemed acceptable for surgical training and digital twin verification by clinicians. FEA further validated these findings, with simulated deformation patterns closely matching experimental observations from the gel phantom and porcine liver specimens (Fig. 4). These results support using K2 and similar formulations in soft tissue simulation studies when ethical or logistical constraints limit access to biological organs. The findings provide critical insights into gel-based liver phantom capabilities for replicating tactile and visual responses of biological tissue, with valuable implications for integrating synthetic organ models in robot-assisted surgical training, preoperative planning, and digital twin frameworks.

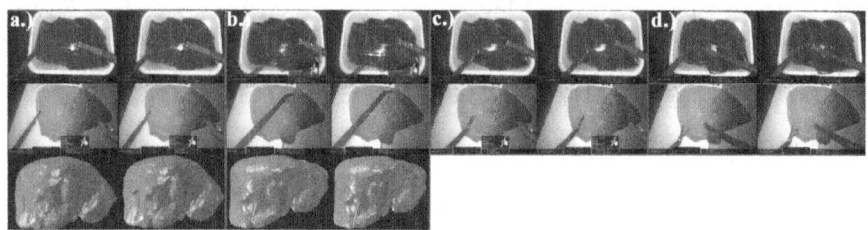

Fig. 4. Instrument actions on different tissues. Top: porcine tissue; middle: gel phantom; bottom: FEM. a) Compression; b) Pulling; c) Pinching; d) Cutting.

3.4 Ultrasound Scanning

Surface scanning assessment demonstrated that tumours embedded within the Konnyaku remained detectable from the surface (Fig. 5). Portable ultrasound examination by hepatobiliary liver surgeon expert at the known tumour location confirmed visibility both visually and through ultrasound video obtained.

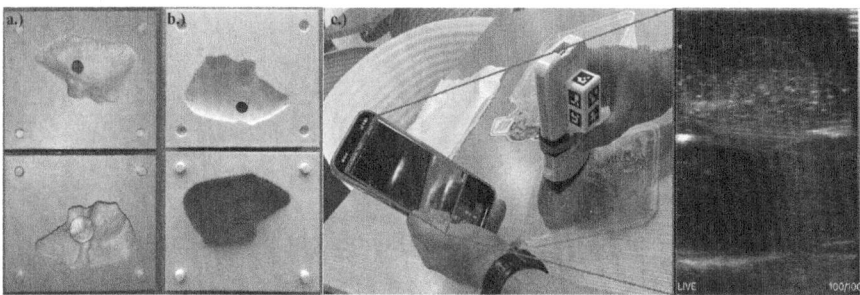

Fig. 5. Ultrasound scanning of the liver phantom. a.) 3D printed mould with a tumor before casting; b.) The settled liver phantom by adding food coloring containing tumor; c.)Ultrasound scanning of the liver phantom by hepatobiliary expert.

3.5 Limitation and Future Work

While this study successfully demonstrated the feasibility of fabricating liver phantoms using food-grade hydrogels for digital twin modeling and surgical training, its primary contribution lies in advancing the VVUQ frameworks, which is a critical foundation for establishing digital twin reliability and clinical deployment in healthcare systems. However, due to several methodological limitations, further research is needed to optimize the digital twin workflow.

The mechanical evaluation was limited to basic tensile and compressive testing, which inadequately captures the complex biomechanical behavior essential for digital twin verification processes. Real liver tissue exhibits sophisticated viscoelastic properties, including time-dependent stress relaxation and anisotropic responses that must be accurately represented in digital twin models to ensure clinical fidelity. These characterization gaps introduce significant uncertainties in computational model parameters, compromising the verification process where numerical solutions must be validated against known analytical or experimental benchmarks. Future work should incorporate comprehensive biomechanical evaluation including viscoelastic profiling through dynamic mechanical analysis, cyclic loading protocols, and multi-scale indentation testing. By providing robust experimental data, enhanced characterization would reduce parametric uncertainties and strengthen verification procedures for digital twin model validation.

Additionally, the reliance on qualitative tactile assessment introduces subjectivity that fundamentally undermines systematic validation of digital twin haptic outputs against physical phantom responses. This limitation reflects critical challenges in establishing quantitative validation metrics for digital twin systems in healthcare, where subjective clinical assessments must be translated into objective validation criteria for regulatory approval. The absence of standardized quantitative metrics impedes uncertainty quantification processes that are essential for establishing confidence intervals in digital twin predictions. Future investigations should establish comprehensive quantitative frameworks incorporating force-displacement mapping, material damping coefficients, and standardized haptic protocols. These metrics would enable rigorous validation of digital twin force feedback systems and support uncertainty quantification through statistical comparison of predicted versus measured responses.

With regard to anatomical and biological reference constraints, the utilization of a single donor-based liver model constrains the validation scope of digital twin systems, as healthcare applications require validation across diverse patient populations with varying anatomical geometries and pathological conditions. This limitation directly impacts uncertainty quantification processes, as digital twin predictions must account for inter-patient variability to provide clinically meaningful confidence intervals. The use of porcine liver as biological reference introduces systematic biases that propagate through the entire VVUQ pipeline, potentially invalidating digital twin systems when deployed in human healthcare contexts. Addressing these limitations requires developing patient-specific validation frameworks incorporating diverse geometries from clinical databases and establishing uncertainty quantification methodologies that account for anatomical variability in digital twin predictions.

Furthermore, evaluation under static conditions omits dynamic physiological elements critical for healthcare digital twin applications, including perfusion dynamics, bleeding responses, and respiratory motion. These omissions introduce model uncertainties that cannot be quantified without incorporating physiological variability into validation protocols. Healthcare digital twins must operate reliably under dynamic clinical conditions, requiring validation frameworks that capture the full spectrum of physiological states encountered during patient care. Future platforms should integrate perfusion systems, bleeding simulation, and physiological motion to enable comprehensive validation of digital twin responses under realistic clinical scenarios.

Current phantoms lack sensing capabilities essential for real-time validation and uncertainty quantification in digital twin systems. Healthcare applications require continuous model updating based on patient-specific data, demanding embedded sensors for real-time validation of digital twin predictions. Future research could embed distributed sensor networks including strain gauges for deformation tracking, pressure sensors for contact force measurement, and fiber-optic sensors for distributed strain sensing while maintaining MRI compatibility. This technological integration would facilitate cyber-physical systems where physical phantom manipulation directly informs computational model refinement, creating dynamic digital twins that evolve based on experimental observations—a critical advancement toward clinically deployable surgical simulation and guidance systems.

These limitations identified reflect broader challenges in surgical simulation and healthcare digital twin development, emphasizing the critical need for interdisciplinary collaboration and standardized validation protocols to ensure clinical deployment readiness.

4 Conclusion

This study successfully developed and validated novel food-grade hydrogel liver phantoms replicating healthy, fatty, and fibrotic tissue properties. Validated by 25 clinicians and comprehensive mechanical testing, these phantoms provide critical physical testbeds for verification, validation, and uncertainty quantification in digital twin frameworks. The phantoms bridge computational models and clinical reality, enabling surgeon-in-the-loop simulations essential for robotic surgery. While limitations exist in mechanical characterization scope and anatomical variation, this work establishes a foundational platform for digital twin development in liver surgery. Future sensor integration and expanded characterization will advance these tools toward clinical implementation, ultimately improving surgical precision and patient outcomes.

Acknowledgments. This study was funded by the NMRC Clinician Investigator Award (CIA) Grant 2023 (Grant ID: CIAINV23jul-0003, Project ID: MOH-001469).

The authors also acknowledge Strength of Materials Lab at the National University of Singapore and the Holomedicine team from the Academic Innovation Office, National University Health System, Singapore.

Disclosure of Interests. The authors have no competing interests to declare that are relevant to the content of this article.

References

1. Ali, S., et al.: An objective comparison of methods for augmented reality in laparoscopic liver resection by preoperative-to-intraoperative image fusion from the miccai2022 challenge. Med. Image Anal. **99**, 103371 (2025)
2. Allard, J., Courtecuisse, H., Faure, F.: Implicit fem solver on GPU for interactive deformation simulation. In: GPU Computing Gems Jade Edition, pp. 281–294. Morgan Kaufmann (2012)
3. Bui, H., Tomar, S., Bordas, S.: Corotational cut finite element method for real-time surgical simulation: Application to needle insertion simulation. Computer Methods in Applied Mechanics and Engineering (2018)
4. Chmarra, M., Hansen, R., Marvik, R., Lango, T.: Multimodal phantom of liver tissue. PLoS ONE **8**(5), e64180 (2013)
5. Collins, J.A., et al.: Improving registration robustness for image-guided liver surgery in a novel human-to-phantom data framework. IEEE Trans. Med. Imaging **36**(7), 1502–1510 (2017)
6. Ding, H., Seenivasan, L., Killeen, B., Cho, S., Unberath, M.: Digital twins as a unifying framework for surgical data science: the enabling role of geometric scene understanding. Artif. Intell. Surg. **4**(3), 109–138 (2024)
7. Elisei, R.C., et al.: Liver phantoms cast in 3D-printed mold for image-guided procedures. Diagnostics **14**(14) (2024)
8. Fu, W., et al.: iPhantom: a framework for automated creation of individualized computational phantoms and its application to CT organ dosimetry. IEEE J. Biomed. Health Inform. **25**(8), 3061–3072 (2021)
9. Han, Z., Dou, Q.: A review on organ deformation modeling approaches for reliable surgical navigation using augmented reality. Comput. Assist. Surg. **29**(1), 2357164 (2024)
10. Heiselman, J.S., et al.: Characterization and correction of intraoperative soft tissue deformation in image-guided laparoscopic liver surgery. J. Med. Imaging **5**(2), 021203 (2018)
11. Lemine, A., Ahmad, Z., Al-Thani, N., Hasan, A., Bhadra, J.: Mechanical properties of human hepatic tissues to develop liver-mimicking phantoms for medical applications. Biomech. Model. Mechanobiol. **23**(2), 373–396 (2024)
12. Mannle, D., et al.: Artificial intelligence directed development of a digital twin to measure soft tissue shift during head and neck surgery. PLoS ONE **18**(8), e0287081 (2023)
13. Mendizabal, A., et al.: Face-based smoothed finite element method for real-time simulation of soft tissue. In: SPIE Medical Imaging. SPIE, Orlando, United States (2017)
14. Peterlik, I., et al.: Fast elastic registration of soft tissues under large deformations. Med. Image Anal. **45**, 24–40 (2018)

15. Peterlik, I., Haouchine, N., Ručka, L., Cotin, S.: Image-driven stochastic identification of boundary conditions for predictive simulation. In: MICCAI. Springer, Québec, Canada (2017)
16. Plantefeve, R., Peterlik, I., Haouchine, N., Cotin, S.: Patient-specific biomechanical modeling for guidance during minimally-invasive hepatic surgery. Ann. Biomed. Eng. **44**(1), 139–153 (2016)
17. Reichard, D., et al.: Projective biomechanical depth matching for soft tissue registration in laparoscopic surgery. Int. J. Comput. Assist. Radiol. Surg. **12**(7), 1101–1110 (2017)
18. Rethy, A., et al.: Anthropomorphic liver phantom with flow for multimodal image-guided liver therapy research and training. Int. J. Comput. Assist. Radiol. Surg. **13**(1), 61–72 (2018)
19. Schüßler, A., Younis, R., Paik, J., Wagner, M., Mathis-Ullrich, F., Kunz, C.: Interactive surgical liver phantom for cholecystectomy training. Curr. Dir. Biomed. Eng. **10**(2), 95–98 (2024)
20. Servin, F., et al.: Simulation of image-guided microwave ablation therapy using a digital twin computational model. IEEE Open J. Eng. Med. Biol. **5**, 107–124 (2024)
21. Servin, F., et al.: Digital twin modeling and machine learning frameworks for forecasting multiple microwave ablation volumes. In: Medical Imaging 2025: Image-Guided Procedures, Robotic Interventions, and Modeling. SPIE (2025)
22. Shi, Y., et al.: Synergistic digital twin and holographic augmented-reality-guided percutaneous puncture of respiratory liver tumor. IEEE Trans. Hum. Mach. Syst. **52**(6), 1364–1374 (2022)
23. Spinczyk, D.: Measuring respiratory motion for supporting the minimally invasive destruction of liver tumors. Sensors **22**(17) (2022)
24. Subramanian, K.: Digital twin for drug discovery and development–the virtual liver. J. Indian Inst. Sci. **100**(4), 653–662 (2020)
25. Valls-Esteve, A., et al.: Patient-specific 3D printed soft models for liver surgical planning and hands-on training. Gels **9**(4) (2023)
26. Wasserthal, J., et al.: TotalSegmentator: robust segmentation of 104 anatomic structures in CT images. Radiol. Artif. Intell. **5**(5), e230024 (2023)
27. Wu, J., Westermann, R., Dick, C.: Real-time haptic cutting of high-resolution soft tissues. In: Medicine Meets Virtual Reality 21. Studies in Health Technology and Informatics, IOS Press (2014)

A Real-Time Digital Twin for Type 1 Diabetes Using Simulation-Based Inference

Trung-Dung Hoang[1,2,3]✉, Alceu Bissoto[1,2,3], Vihangkumar V. Naik[1,2,3], Tim Flühmann[1,2,3], Artemii Shlychkov[1,2,3,4], Jose Garcia-Tirado[1,2,3], and Lisa M. Koch[1,2,3]✉

[1] University of Bern, Bern, Switzerland
{trung.hoang,lisa.koch}@unibe.ch
[2] Department of Diabetes, Endocrinology, Nutritional Medicine and Metabolism, Inselspital, Bern University Hospital, University of Bern, Bern, Switzerland
[3] Diabetes Center Berne, Bern, Switzerland
[4] University of Tübingen, Tübingen, Germany

Abstract. Accurately estimating parameters of physiological models is essential to achieving reliable digital twins. For Type 1 Diabetes, this is particularly challenging due to the complexity of glucoseinsulin interactions. Traditional methods based on Markov Chain Monte Carlo struggle with high-dimensional parameter spaces and fit parameters from scratch at inference time, making them slow and computationally expensive. In this study, we propose a Simulation-Based Inference approach based on Neural Posterior Estimation to efficiently capture the complex relationships between meal intake, insulin, and glucose level, providing faster, amortized inference. Our experiments demonstrate that SBI not only outperforms traditional methods in parameter estimation but also generalizes better to unseen conditions, offering real-time posterior inference with reliable uncertainty quantification. Code available at https://github.com/mlm-lab-research/SBI_T1D.

1 Introduction

Type 1 diabetes (T1D) affects more than 9 million people worldwide [1]. T1D is an autoimmune condition resulting in an absolute deficiency in insulin, a hormone essential for allowing glucose to enter cells and produce energy. Individuals with T1D therefore require frequent insulin injections and close monitoring of their glucose levels, e.g., with wearable continuous glucose monitoring (CGM) devices. Maintaining healthy glucose levels through carefully administered insulin is crucial: too much insulin can lead to life-threatening hypoglycaemia, while too little insulin results in increased blood glucose levels, which can cause many serious long-term micro- and macrovascular complications.

The dynamics between glucose, meal intake, and insulin over time can be described by complex physiological models consisting of systems of differential equations [2–4], where model parameters (e.g., insulin sensitivity) can be highly patient-specific. Identifying patient-specific parameters from observed data allows creating a digital twin (DT) of the individual's metabolic system. This could improve treatment planning, prediction, and real-time adaptation, where the DT is continuously updated with patient data, allowing the system to better respond to daily variations in meals, insulin dosing, and insulin sensitivity [5].

Despite the promise of DTs, parameter estimation in T1D models is a challenging inverse problem due to complex, nonlinear relationships among system components. The current gold standards (e.g., Markov Chain Monte Carlo (MCMC)) [6–9] attempt to directly fit the physiological model to individual data. This means they require a separate optimization or sampling process for each new observation (non-amortized), making them computationally expensive, limited to low-dimensional problems, and unsuitable for real-time applications. Furthermore, an often overlooked challenge is inferring the system's initial conditions. Many current approaches are limited to a steady-state initialisation. In the case of T1D, this means no meals or insulin could be administered for at least four hours before the observation window starts. If the assumption is not met, the inferred parameters may overcompensate for the wrong initial conditions, resulting in inaccurate parameter estimates and inadequate use of the DT.

To address these challenges, Simulation-Based Inference (SBI) offers a more flexible and scalable approach that uses neural networks for probabilistic inference [10]. SBI has been successfully applied to inverse problems in different fields, such as astrophysics [11–13] and cardiovascular simulations [14]. SBI requires only data simulated from physiological models to train the neural network, enabling it to work directly with any simulator. Furthermore, by leveraging Normalizing Flows [15] in particular the Neural Posterior Estimation (NPE) framework [16,17], SBI can efficiently model complex, high-dimensional posterior distributions, allowing inference over a larger set of parameters.

In this work, we present the first application of SBI to CGM data and T1D models. Our contributions are summarized as follows:

- First application of SBI to T1D models: We demonstrate how SBI can be effectively applied to CGM data and physiological models of T1D.
- Improved parameter estimation: We compare our approach to the previous state-of-the-art method [6] and show that SBI offers a powerful method for parameter inference.
- Inference of initial conditions: We show how SBI can be naturally extended to estimate initial conditions of physiological models, such that our proposed approach works reliably in more realistic scenarios beyond steady-state conditions.
- Computational efficiency: Once trained, the neural network can rapidly infer parameters for new patients without running a separate optimization or sampling routine.

2 Background

2.1 Physiological T1D Models

Physiological T1D models aim to mathematically represent how the body regulates blood glucose through the dynamics of insulin and glucose in people with T1D. In this work, we use a simplified version [6] of the UVA/Padova simulator [18], a widely accepted and clinically validated physiological model of T1D. This simplification is a standard in the area, enabling fair comparisons with our baselines while preserving a challenging setup that accurately reflects glucose-insulin dynamics. The model is divided into four main subsystems:

Subcutaneous insulin absorption models how delivered insulin I reaches the plasma after injection. It first transitions to non-monomeric insulin I_{sc1}, then to monomeric insulin I_{sc2}, and finally to plasma insulin I_p:

$$\begin{cases} \dot{I}_{sc1}(t) = -k_d \cdot I_{sc1}(t) + I(t - \beta)/V_I \\ \dot{I}_{sc2}(t) = k_d \cdot I_{sc1}(t) - k_{a2} \cdot I_{sc2}(t) \\ \dot{I}_p(t) = k_{a2} \cdot I_{sc2}(t) - k_e \cdot I_p(t) \end{cases}$$

Here, k_d is the transition rate from non-monomeric to monomeric insulin, k_{a2} is the absorption rate into plasma, k_e is the clearance rate, V_I is the insulin distribution volume, and β is the insulin appearance delay.

Oral glucose absorption models how ingested carbohydrates CHO transition through the digestive system. Glucose enters the stomach as solid state (Q_{sto1}), converts to liquid state (Q_{sto2}), moves to the intestine (Q_{gut}), and finally appears in plasma as Ra (rate of glucose appearance):

$$\begin{cases} \dot{Q}_{sto1}(t) = -k_{empt} \cdot Q_{sto1}(t) + CHO(t) \\ \dot{Q}_{sto2}(t) = k_{empt} \cdot Q_{sto1}(t) - k_{empt} \cdot Q_{sto2}(t) \\ \dot{Q}_{gut}(t) = k_{empt} \cdot Q_{sto2}(t) - k_{abs} \cdot Q_{gut}(t) \\ Ra(t) = f \cdot k_{abs} \cdot Q_{gut}(t) \end{cases}$$

Here, k_{empt} is the gastric emptying rate, and k_{abs} and f are the intestinal absorption rate and fraction of glucose absorption through the intestinal wall, respectively.

Glucose-insulin kinetics models how glucose and insulin interact in the body. Plasma glucose G is affected by insulin action X and glucose appearance Ra. Insulin action X depends on plasma insulin I_p, and interstitial glucose IG reflects glucose transport between plasma and the interstitium:

$$\begin{cases} \dot{G}(t) = -\rho(G) \cdot X(t) \cdot G(t) - SG \cdot [G(t) - G_b] + Ra(t)/V_G \\ \dot{X}(t) = -p_2 \cdot X(t) + p_2 \cdot SI \cdot (I_p(t) - I_{pb}) \\ \dot{IG}(t) = -(IG(t) - G(t))/\alpha \end{cases}$$

Here, SG is the fractional glucose effectiveness, quantifying how well glucose regulates itself by stimulating its uptake and suppress its production to maintain basal glucose levels G_b; V_G is the glucose distribution volume; SI is insulin sensitivity, measuring how effectively insulin lowers blood glucose; I_{pb} is the basal insulin concentration; p_2 is the insulin action decay rate, and α is the plasma-to-interstitium transport delay. The function $\rho(G)$ [6] increases insulin action X under hypoglycemia to better capture glucose dynamics in the low-glucose range.

CGM sensor error models the error introduced by the CGM sensor:

$$CGM(t) = [(a_0 + a_1 \cdot t + a_2 \cdot t^2) \cdot IG(t) + b_0] + v(t)$$

where a_0, a_1, a_2, b_0 are sensor coefficients; $v(t) \sim N(0, \epsilon_v)$ is white noise.

Given meal inputs $CHO(t)$, insulin inputs $I(t)$, initial state values, and model parameters, the simulator generates CGM readings $CGM(t)$. Some parameters, such as SI, are highly patient-specific [19]. Our goal is to estimate these individual-specific parameters from observed inputs and outputs.

2.2 Simulation-Based Inference (SBI)

Traditional Bayesian inference computes the posterior distribution of parameters $\boldsymbol{\theta}$ given observed data $\boldsymbol{y}$ as $p(\boldsymbol{\theta} \mid \boldsymbol{y}) \propto p(\boldsymbol{y} \mid \boldsymbol{\theta}) \cdot p(\boldsymbol{\theta})$, where $p(\boldsymbol{y} \mid \boldsymbol{\theta})$ is the likelihood and $p(\boldsymbol{\theta})$ is the prior. In many scientific applications, including physiological T1D models, even though we can simulate $\boldsymbol{y}$ from the model given any $\boldsymbol{\theta}$, the likelihood function $p(\boldsymbol{y} \mid \boldsymbol{\theta})$ is intractable - we cannot evaluate it explicitly or compute its value efficiently. This happens when the simulator involves stochasticity, latent states, or complex numerical solvers. As a result, inference methods such as MCMC often assume a tractable form (e.g., a Gaussian likelihood), which may fail to capture the true complexity of the models.

Simulation-Based Inference (SBI) [10], also known as likelihood-free inference, overcomes this by relying only on a simulator that can generate synthetic data $\tilde{\boldsymbol{y}} \sim p(\boldsymbol{y} \mid \boldsymbol{\theta})$ for any chosen parameter configuration $\boldsymbol{\theta}$. The goal is to learn the posterior distribution $p(\boldsymbol{\theta} \mid \boldsymbol{y})$ directly from these samples, without requiring an explicit likelihood function. A powerful SBI approach is Neural Posterior Estimation (NPE) [16,17], which uses Conditional normalizing flows [17] to approximate the posterior. A normalizing flow f_ϕ [15] is an invertible neural network that transforms a simple distribution p_Z (e.g., a standard Gaussian) into a more complex one. In the conditional setting, the transformation depends on $\boldsymbol{y}$:

$$\boldsymbol{\theta} = f_\phi(\boldsymbol{z}; \boldsymbol{y}), \quad \boldsymbol{z} \sim p_Z(\boldsymbol{z})$$

The resulting density over $\boldsymbol{\theta}$ is given by the change-of-variables formula:

$$q_\phi(\boldsymbol{\theta} \mid \boldsymbol{y}) = p_Z(\boldsymbol{z}) \left| \det\left(\frac{\partial f_\phi^{-1}(\boldsymbol{\theta}; \boldsymbol{y})}{\partial \boldsymbol{\theta}} \right) \right|$$

which is used to approximate $p(\boldsymbol{\theta} \mid \boldsymbol{y})$.

To train the flow, we maximize the conditional density $q_\phi(\boldsymbol{\theta}^{(i)} \mid \tilde{\boldsymbol{y}}^{(i)})$ for simulated pairs $(\boldsymbol{\theta}^{(i)}, \tilde{\boldsymbol{y}}^{(i)})$. At inference, given observed $\boldsymbol{y}$, we can sample multiple

$z \sim p_Z(z)$ and apply the learned f_ϕ to obtain $\boldsymbol{\theta} = f_\phi(z; y)$ from the approximate posterior. This approach enables efficient parameter estimation in complex physiological models without any assumption about the likelihood function.

3 SBI for T1D Model Identification

3.1 Parameters and Prior Distribution

We apply SBI to estimate parameters of the T1D model described in Sect. 2.1. Following the state-of-the-art ReplayBG baseline, which uses MCMC as its backend [6], we focus on estimating a subset of eight physiological parameters, while keeping the remaining parameters fixed at population-level constants:

$$\boldsymbol{\theta} = [G_b, SG, p_2, k_{a2}, k_d, k_{empt}, SI, k_{abs}]$$

We propose to treat the initial conditions as model parameters, enabling joint inference with physiological parameters via SBI. This helps avoid compensatory effects and bias caused by misspecified initial states. We consider the following nine initial state variables:

$$\mathbf{x}_0 = [G(0), I_{sc1}(0), I_{sc2}(0), I_p(0), Q_{sto1}(0), Q_{sto2}(0), Q_{gut}(0), X(0), IG(0)]$$

The full vector to be inferred is thus $\hat{\boldsymbol{\theta}} = [\boldsymbol{\theta}, \mathbf{x}_0] \in \mathbb{R}^{17}$. The observation vector $y \in \mathbb{R}^{264}$ consists of CGM readings recorded every 5 min over 22 h, as in the ReplayBG implementation [6]. The prior over $\hat{\boldsymbol{\theta}}$ is constructed as $p(\hat{\boldsymbol{\theta}}) = p(\boldsymbol{\theta}, \mathbf{x}_0) = p(\boldsymbol{\theta}) \cdot p(\mathbf{x}_0 \mid \boldsymbol{\theta})$. The marginal prior $p(\boldsymbol{\theta})$ follows the ReplayBG configuration. Since $p(\mathbf{x}_0 \mid \boldsymbol{\theta})$ has no closed-form, we propose sampling $\boldsymbol{x}_0$ by simulating 44 h from a known steady state using the given $\boldsymbol{\theta}$ and extracting a random 22-hour window as y. The initial nine state values from this shifted observation window are used as $\mathbf{x}_0$.

3.2 Training Details

To generate training data, we draw parameter vectors $\hat{\boldsymbol{\theta}}$ and, for each sample, simulate CGM observations using the T1D model with a fixed meal and insulin profile. We use rejection sampling and retain only the simulations with CGM outputs in the range [40, 400] mg/dL until 5,000 valid samples are collected.

For the posterior estimator f_ϕ, we use the implementation from the sbi library [20]. Specifically, we employ a Masked Autoregressive Flow (MAF) [21] composed of Masked Affine Autoregressive Transform layers, each conditioned on the observation y. We train f_ϕ with the following configuration: batch size of 200, learning rate of 5×10^{-4}, gradient norm clipped at 5.0, and validation fraction of 10%. Early stopping is triggered after 20 epochs of no improvement.

All the computation was done using a NVIDIA GeForce RTX 4090 GPU (24 GB VRAM) and 4 CPU cores with 64 GB of RAM. The system ran Rocky Linux 9.5 (Blue Onyx) with CUDA 12.4 and NVIDIA driver version 550.163.01.

3.3 Evaluation and Baselines

We evaluate our T1D digital twin on 50 simulated CGM trajectories, each generated from a different parameter vector drawn from the prior distribution. For each CGM trajectory, we use SBI to infer 1,000 posterior samples. We then assess the quality of these posterior estimates by comparing them against the ground truth parameters using four metrics.

The **absolute error (Abs Error)** and **relative error (Rel Error)** measure the absolute and relative difference between the median of the 1,000 posterior samples and the true value, respectively, while the **mean absolute deviation (MAD)** captures the average absolute difference across all posterior samples. The **coverage** metric indicates whether the true values lie within the 2.5th and 97.5th percentiles of the posterior samples. We also evaluate our approach's ability to reconstruct the observed CGM trajectory using **Mean Absolute Relative Difference (MARD)** and **Root Mean Square Error (RMSE)** between the observed and the reconstructed signal, averaged over the 50 test samples.

We compare our approach to two baselines: ReplayBG [6] and a Maximum a posteriori (MAP) approach using parameter search algorithms to find a point estimate which maximizes the posterior $p(\boldsymbol{\theta}|\boldsymbol{y})$. MCMC in ReplayBG is configured with 10,000 burn-in and 5,000 main steps. For the MAP approach, we use the implementation provided in [6].

4 Results

4.1 SBI Improves Parameter Estimation in T1D Digital Twins

SBI consistently led to the lowest parameter estimation errors (Fig. 1 a,b,c). For basal glucose (G_b), a crucial parameter for T1D management, SBI attained an error of 5.04 mg/dl (4.32% relative error) and a MAD of 7.19 mg/dl, compared to ReplayBG's 19.12 mg/dl (16.24%) and 19.15 mg/dl, respectively. The smaller errors of SBI could be attributed to its ability to learn complex distributions and estimate the initial conditions, whereas ReplayBG fixed the initial conditions at steady state and had to compensate the parameters to fit the observation. By jointly estimating initial conditions and parameters, SBI could more accurately capture the true system dynamics.

The uncertainty quantification obtained through SBI's full posterior estimates led to large coverage for all parameters (Fig. 1 d), with an average coverage of 96.5% within the 95% credible interval, indicating well-calibrated uncertainty. In contrast, the MCMC posteriors were consistently overconfident, which can cover the true values in only 23.25% of cases on average. These results are further supported qualitatively in Fig. 2, where posteriors of SBI covered the true parameters best. The MAP baseline provided only a point estimate without uncertainty quantification, making coverage calculation inapplicable.

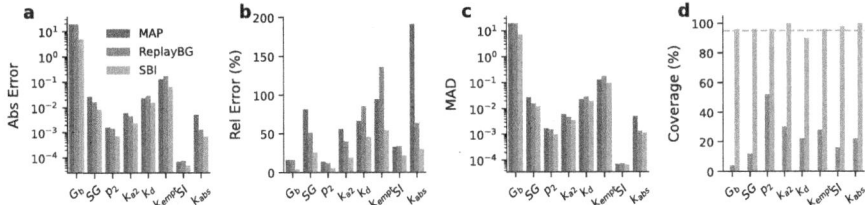

Fig. 1. SBI outperforms baseline methods: **a-c)** Parameter estimation error and **d)** Coverage of inferred parameter posteriors (dashed line shows 95% coverage). MAP (red) produces single estimations and can not cover the ground-truth.

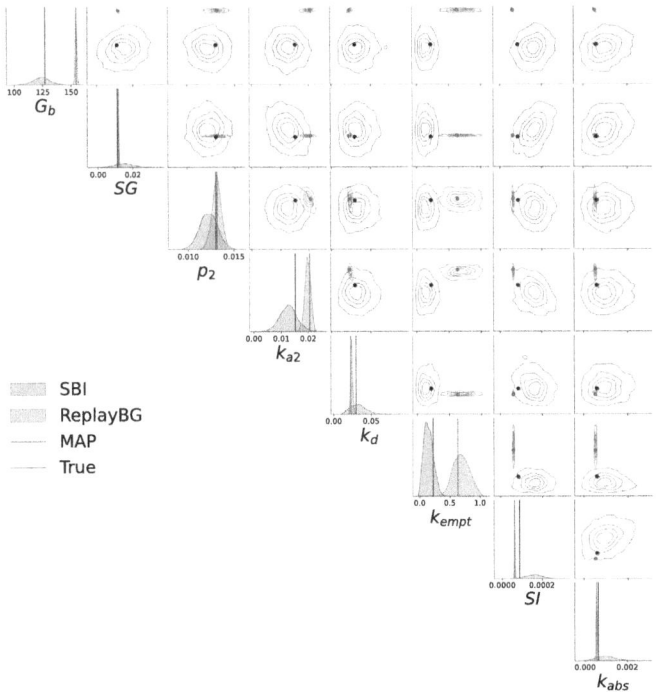

Fig. 2. Visualization of posterior distributions of a single test case. SBI posteriors (blue) cover the true value (black) well, while MCMC posteriors (green) tend to be narrower and overconfident around the MAP estimate (red). See more examples in Appendix A (Color figure online).

4.2 What-if Scenario Simulation: SBI Generalises to Unseen Scenarios

A key motivation for using DTs is to explore counterfactual or hypothetical scenarios that cannot be directly observed in real-world data. To evaluate generalization ability, we considered three settings where CGM signals were reconstructed from inferred parameters: (1) **In-sample reconstruction** - simulation using the same meal profile as in the observation; (2) **Out-of-sample: next day** - simulation extended to the next day; (3) **Out-of-sample: altered meals** - simulation with the meal profile modified. For SBI and ReplayBG, we used the median CGM signal simulated from 1,000 posterior samples as a replay signal, and for MAP, we used a single CGM signal simulated from the point estimate.

ReplayBG outperformed SBI in the in-sample setting, as they are designed to fit the observed data closely (left columns in Table 1). However, in out-of-sample scenarios, SBI consistently outperformed both baselines, achieving the lowest MARD and RMSE (middle and right columns in Table 1). Figure 3 shows an example test case. While MAP and ReplayBG accurately reproduced the CGM on the first day, their replay signals deviated considerably on the second day. In contrast, SBI continued to follow the ground-truth trajectory thanks to SBI's more accurate estimation of the physiological parameters, which is crucial for generalizing to unseen conditions.

Table 1. Average MARD (%) and RMSE (mg/dL) across three settings.

	in-sample reconstruction		out-of-sample: next day		out-of-sample: altered meals	
	MARD	RMSE	MARD	RMSE	MARD	RMSE
MAP	14.36 ± 6.00	17.30 ± 5.43	22.29 ± 15.47	23.81 ± 12.12	18.11 ± 7.84	21.97 ± 8.45
ReplayBG	**6.73 ± 2.82**	**8.31 ± 2.20**	20.20 ± 16.69	20.76 ± 10.83	15.05 ± 6.53	17.83 ± 7.40
SBI	12.60 ± 6.19	15.12 ± 6.79	**14.84 ± 8.84**	**16.09 ± 7.54**	**12.68 ± 6.42**	**15.56 ± 8.60**

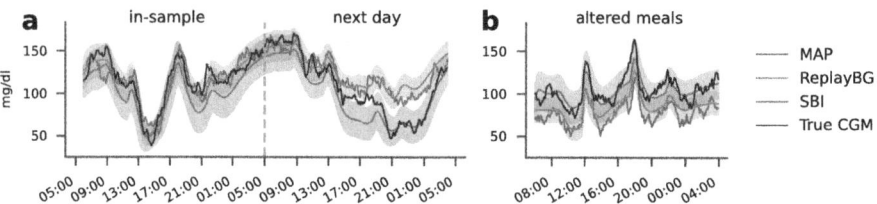

Fig. 3. CGM reconstructions (colors depict different methods) for an example CGM signal (black). **a)** All methods matched the observed data on the first day (Setting 1), but only SBI accurately predicted the next day (Setting 2). **b)** SBI reconstructed the true CGM better than the baselines (Setting 3).

4.3 SBI is Suitable for Real-Time Inference

Table 2 highlights the efficiency of SBI in terms of inference speed and overall computational cost. Inference time excluded the time to load the model or data. Inference using SBI took 3.36 s on average, outperforming both baselines by a large margin. While MAP inference completed in less than a minute, it often yielded large errors. It also lacks uncertainty quantification, which is crucial for facilitating trust in the inference results. ReplayBG can produce posterior samples but required approximately 45 min per sample. Although SBI needed 161.2 s to generate training data and train the model, this one-time cost enables reuse without retraining, providing amortized inference. This makes SBI a great choice for tasks that need frequent or real-time inference, such as DTs, where they must quickly adapt to changes in the real system.

Table 2. Inference and training time comparison across methods.

Method	Inference Time (s)	Training Time (s)
MAP	51.46 ± 13.98	0
ReplayBG	2767.04 ± 258.36	0
SBI	3.36 ± 0.52	161.2

5 Discussion

In this work, we propose using SBI for T1D digital twins. SBI is an efficient approach that enables accurate, efficient, and amortized parameter inference. Our extension of SBI to infer initial conditions enables it to operate beyond steady-state conditions. This makes it particularly well-suited for complex physiological systems where traditional methods often rely on those restrictive assumptions. Our experiments demonstrate that SBI delivers accurate posterior estimates and significantly improves the quality of simulated CGM trajectories. Its fast, amortized inference opens the door for real-time DTs, which remain a challenge for existing methods.

While promising, SBI still faces several limitations. Since it relies on simulated data, the quality of its posterior approximation depends heavily on the coverage of that data. The inferred posterior may be less accurate or overly confident in regions that are underrepresented in the training set. This can happen when the simulation budget is limited or the prior distribution is poorly specified. Furthermore, SBI models typically require retraining if the prior or simulator is modified.

To further enhance the robustness of SBI, for future work, we plan to simulate a broader range of scenarios, such as varying meal and insulin profiles, to enrich the training data. Beyond glucose modeling, SBI offers opportunities for DTs in other domains, such as cardiac dynamics, making it a truly transformative tool in computational physiology.

Acknowledgments. This project was supported by the Diabetes Center Berne and strategic funding of the medical faculty of the University of Bern. Calculations were performed on UBELIX, the HPC cluster at the University of Bern.

Disclosure of Interests. The authors have no competing interests to declare.

A Additional Figures

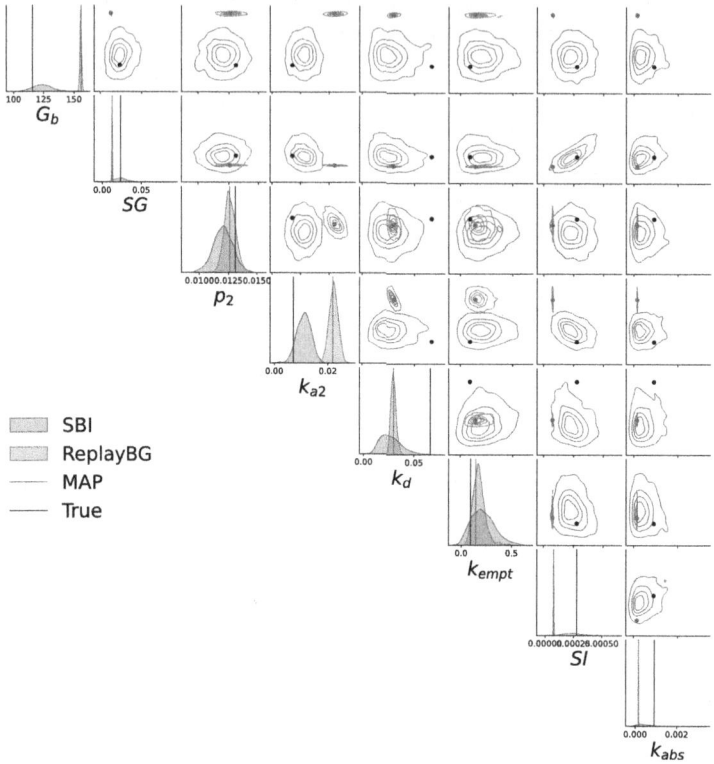

Fig. 4. Visualization of posterior distributions for a second test case.

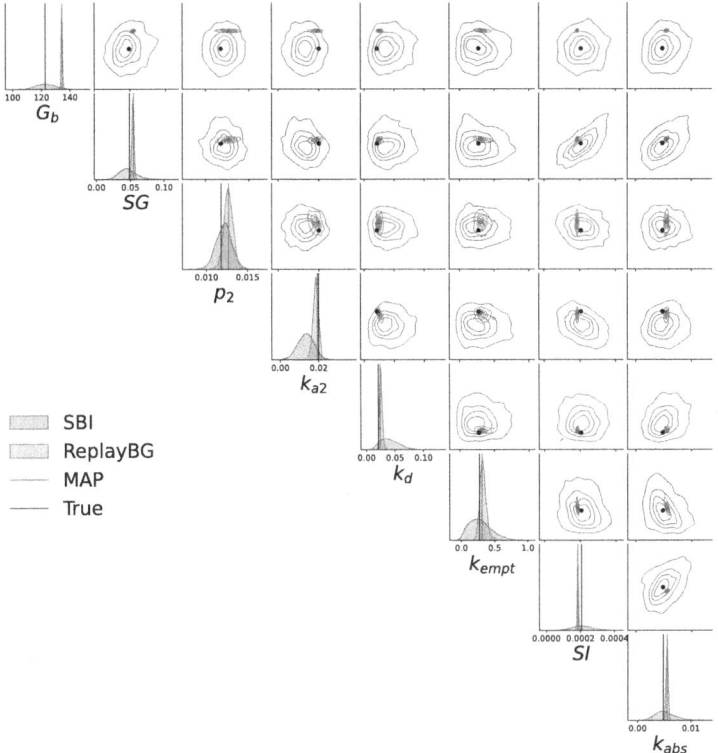

Fig. 5. Posterior visualization for a third test case.

References

1. Global impact dashboard (2025). https://dashboard.t1dindex.org/global-impact. Accessed 01 June 2025
2. Dalla Man, C., Rizza, R.A., Cobelli, C.: Meal simulation model of the glucose-insulin system. IEEE Trans. Biomed. Eng. **54**(10), 1740–1749 (2007)
3. Foster, R.O., Soeldner, J.S., Tan, M.H., Guyton, J.R.: Short term glucose homeostasis in man: A systems dynamics model. J. Dyn. Syst. Measur. Control **95**(3), 308–314 (1973). https://doi.org/10.1115/1.3426720
4. Bergman, R.N., Ider, Y.Z., Bowden, C.R., Cobelli, C.: Quantitative estimation of insulin sensitivity. Am. J. Physiol. Endocrinol. Metab. **236**(6), E667 (1979)
5. Kovatchev, B., et al.: Human-machine co-adaptation to automated insulin delivery: a randomised clinical trial using digital twin technology. NPJ Dig. Med. **8** (2025). https://doi.org/10.1038/s41746-025-01679-y
6. Cappon, G., Vettoretti, M., Sparacino, G., Favero, S., Facchinetti, A.: ReplayBG: a digital twin-based methodology to identify a personalized model from type 1 diabetes data and simulate glucose concentrations to assess alternative therapies. IEEE Trans. Biomed. Eng. **70**(11), 3227–3238 (2023)

7. Colmegna, P., Wang, K., Garcia-Tirado, J., Breton, M.D.: Mapping data to virtual patients in type 1 diabetes. Control Eng. Pract. **103**, 104605 (2020). https://doi.org/10.1016/j.conengprac.2020.104605
8. Haidar, A., Wilinska, M.E., Graveston, J.A., Hovorka, R.: Stochastic virtual population of subjects with type 1 diabetes for the assessment of closed-loop glucose controllers. IEEE Trans. Biomed. Eng. **60**(12), 3524–3533 (2013)
9. Visentin, R., Dalla Man, C., Cobelli, C.: One-day Bayesian cloning of type 1 diabetes subjects: toward a single-day UVA/PADOVA type 1 diabetes simulator. IEEE Trans. Biomed. Eng. **63**(11), 2416–2424 (2016)
10. Cranmer, K., Brehmer, J., Louppe, G.: The frontier of simulation-based inference. Proc. Nat. Acad. Sci. **117**(48), 30055–30062 (2020). https://doi.org/10.1073/pnas.1912789117
11. Dax, M., Green, S.R., Gair, J., Macke, J.H., Buonanno, A., Schölkopf, B.: Real-time gravitational wave science with neural posterior estimation. Phys. Rev. Lett. **127**, 241103 (2021). https://doi.org/10.1103/PhysRevLett.127.241103
12. Bhardwaj, U., Alvey, J., Miller, B.K., Nissanke, S., Weniger, C.: Peregrine: Sequential simulation-based inference for gravitational wave signals (2024). https://arxiv.org/abs/2304.02035
13. Mishra-Sharma, S., Cranmer, K.: Neural simulation-based inference approach for characterizing the galactic center γ-ray excess. Phys. Rev. D **105**(6) (2022). https://doi.org/10.1103/physrevd.105.063017
14. Wehenkel, A., et al.: Simulation-based inference for cardiovascular models. In: NeurIPS Workshop (2024). https://arxiv.org/abs/2307.13918
15. Papamakarios, G., Nalisnick, E., Rezende, D.J., Mohamed, S., Lakshminarayanan, B.: Normalizing flows for probabilistic modeling and inference. J. Mach. Learn. Res. **22**(1) (2021)
16. Lueckmann, J.M., et al.: Flexible statistical inference for mechanistic models of neural dynamics. In: Guyon, I., Luxburg, U.V., Bengio, S., Wallach, H., Fergus, R., Vishwanathan, S., Garnett, R. (eds.) Advances in Neural Information Processing Systems. vol. 30. Curran Associates, Inc. (2017). https://proceedings.neurips.cc/paper_files/paper/2017/file/addfa9b7e234254d26e9c7f2af1005cb-Paper.pdf
17. Papamakarios, G., Murray, I.: Fast epsilon -free inference of simulation models with Bayesian conditional density estimation. In: Lee, D., Sugiyama, M., Luxburg, U., Guyon, I., Garnett, R. (eds.) Advances in Neural Information Processing Systems. vol. 29. Curran Associates, Inc. (2016). https://proceedings.neurips.cc/paper_files/paper/2016/file/6aca97005c68f1206823815f66102863-Paper.pdf
18. Visentin, R., et al.: The UVA/PADOVA type 1 diabetes simulator goes from single meal to single day. J. Diabetes Sci. Technol. **12**(2), 273–281 (2018)
19. Thomsen, C., Storm, H., Christiansen, C., Rasmussen, O.W., Larsen, M.K., Hermansen, K.: The day-to-day variation in insulin sensitivity in non-insulin-dependent diabetes mellitus patients assessed by the hyperinsulinemic-euglycemic clamp method. Metabolism **46**(4), 374–376 (1997). https://doi.org/10.1016/S0026-0495(97)90050-0
20. Tejero-Cantero, A., et al.: sbi: A toolkit for simulation-based inference. J. Open Source Softw. **5**(52), 2505 (2020). https://doi.org/10.21105/joss.02505
21. Papamakarios, G., Pavlakou, T., Murray, I.: Masked autoregressive flow for density estimation. In: Proceedings of the 31st International Conference on Neural Information Processing Systems, pp. 2335–2344. NIPS'17, Curran Associates Inc., Red Hook, NY, USA (2017)

Retrospective Evaluation of a Patient-Specific Liver Digital Twin to Predict Thermal Ablation Outcomes in HCC

Chloé Audigier[1(✉)], Felix Meister[2], Fouad Georges Akkari[3], Andrea Tonglet[3], Oliver Frings[2], and Rafael Duran[3]

[1] Siemens Healthineers, Digital Technologies and Innovation, Lausanne, Switzerland
[2] Siemens Healthineers, Digital Technologies and Innovation, Erlangen, Germany
[3] Department of Radiology and Interventional Radiology, Lausanne University Hospital, Lausanne, Switzerland

Abstract. **Liver digital twins** are computer models representing anatomy and physiology in an individualized, accurate and virtual way. In interventional radiology, these models could enable the simulation of several therapeutic strategies before the actual procedure. This is particularly valuable for **Radiofrequency ablation (RFA)**, a well-established minimally invasive treatment for hepatic tumors, where predicting the induced ablation zone remains challenging. Since inaccurate predictions can lead to incomplete treatments or unintended damage to surrounding tissue, a liver digital twin, capable of precisely estimating the expected ablation volume and shape, could drastically improve the ablation outcome.

Although such computational models of RFA have been proposed, they often depend heavily on manual processing of clinical data and are typically evaluated on a few carefully selected cases. In this paper, we present **a retrospective clinical evaluation** of a **fully automatic patient-specific digital twin**, capable of modeling the physical mechanisms involved in RFA of hepatocellular carcinoma (HCC). Our approach integrates clinical information extracted from CT images, from which level set representations of the liver anatomy and vasculatures are automatically extracted. The Lattice Boltzmann Method is used to estimate the temperature propagation and the resulting ablation zone subject to the underlying heat diffusion process. We evaluated our framework on a high-quality dataset comprising 26 patients. The results yield promising correlation between predicted and actual ablation volume (median Relative Volume Difference of 0.37 (IQ: 0.21, 0.51)) and highlight the superiority of our model compared to relying solely on the expected ablation zones as reported in the manufacturer's instructions for use.

Keywords: Liver Digital Twin · Computational Modelling · RFA · Clinical Evaluation · HCC

1 Introduction

Percutaneous **Radiofrequency ablation (RFA)** is a well-established minimally invasive and effective treatment for small (diameter < 3 cm) primary and secondary malignant hepatic tumors [1–3]. The procedure consists of the insertion of a needle to deliver localized heat directly targeting the tumor and causing tissue coagulative necrosis. The ablation is more likely to be successful if a safety margin around the tumor of at least 5 mm is achieved to avoid residual tumor or recurrence [4, 5]. To ensure adequate ablation margin, it is imperative to integrate a predictive model of the expected thermal ablation, parameterized by the position and operational settings of the ablation probe. Inaccurate predictions can lead to incomplete treatments and in turn, an increased risk of recurrence and/or unintended damage to surrounding healthy tissue [6]. However, obtaining a precise estimation of the final ablation volume and shape within hepatic tissue remains a complex and unsolved challenge due to the nonlinear and patient-specific nature of thermal propagation dynamics. While RFA needle manufacturers provide expected ablation zone dimensions for specific ablation settings, these are typically simplified spherical or ellipsoidal approximations derived from ex vivo preclinical data and therefore cannot reliably be translated to patients. Several patient-specific factors such as the state of the liver parenchyma (fibrosis, cirrhotic tissue, etc.) [7, 8], the blood perfusion, the vicinity of the tumor to large blood vessels [9, 10] influence the ablation outcome. In particular, the blood flow in the liver exhibits a protective effect, named the heat sink effect, [11, 12] which is particularly significant if large vessels (diameter > 5mm) are close to the ablation zone. This effect can reduce the efficacy of the thermal ablation and cause high rates of local recurrence [13, 14]. Additionally, when tumors have irregular shapes, a thorough three-dimensional correlation between the expected ablation zone and the tumor location becomes critical. This further complicates the planning process, requiring careful consideration of needle placement and thermal spread to ensure complete tumor coverage.

Liver digital twins are personalized computer models representing the liver of a specific patient in an individualized, accurate, and virtual way. In the context of thermal ablation, liver digital twins combine individual anatomical and physiological parameters, like vasculatures and hepatic perfusion, with numerical methods simulating the heat diffusion in biological tissue to estimate the resulting ablation zone [15–18]. These models may enable clinicians to simulate various ablation strategies virtually before performing the actual procedure on the real patient [19]; thereby ensuring optimal needle placement and ablation configurations to achieve adequate treatment margins.

Although several digital twins have been proposed for RFA, they often depend on manual processing of medical images and additional clinical data [20, 21] and are typically evaluated on a few carefully selected preclinical [22] or clinical cases [23]. In [20], the evaluation is performed on 51 hepatic lesions without distinction between hepatocellular carcinoma (HCC) and other malignant tumors. However, since HCC usually develops in diseased liver (generally cirrhotic) which has a different tissue architecture and composition compared to healthy liver tissue, it is important to consider this distinction.

In this paper, we present **a retrospective clinical evaluation** focusing on **HCC patients** of **a fully automatic** and **patient-specific digital twin** simulating the physical mechanisms involved in RFA. Our approach integrates the patient ablation duration as well as clinical information extracted from CT images, from which level set representations of the liver anatomy and vasculatures are automatically extracted. A heat diffusion model is then solved using the Lattice Boltzmann Method to predict the extent of the ablation zone [16]. The aims of the study are 1) to demonstrate the feasibility of our fully automatic liver digital twin to predict ablation zone, 2) to evaluate the predictive power of our model on HCC patients and, 3) to compare the predictions against the expected ablation zones provided by the probe manufacturer.

2 Materials and Methods

2.1 Clinical Protocol

We performed a retrospective review, approved by the ethics committee, of our patient database for HCC cases treated with RFA and included patients where a single tumor was treated with a single RFA needle and a single impact, who had a CT-guided RFA, and who had undergone multiphasic enhanced CT before, during, and after the thermal ablation procedure. Those multiphasic enhanced CTs had to show the initial tumor, the final needle position, and the final ablation zone for the patient to be accepted.

Ablation Protocol. All patients were presented at a multidisciplinary tumor liver board and gave consent for the procedure. Patients underwent RFA under general anesthesia with jet ventilation and CT guidance using a dedicated interventional CT suite.

Pre-operative Imaging: Imaging was performed during jet ventilation with arms placed above the head and consisted of a nonenhanced scan followed by contrast-enhanced CT in the arterial, and portal phases.

Ablation Planning and Intervention: The pre-operative CT data was used to localize the tumor and plan the electrode trajectory. An interventional radiologist (IR) inserted the needle percutaneously targeting the tumor. Intermittent imaging was obtained until adequate needle positioning was obtained. A nonenhanced CT was acquired to verify the needle placement.

The needle used was the Cool-Tip™ RF Ablation Cluster Electrode (Medtronic, Dublin, Ireland) with an exposure of 2.5 cm, a length of 10, 15 or 20 cm (depending on the tumor depth and patient morphology), with an ablation protocol initially set at 12 min but with 3 roll-offs. As the RF electrode is impedance-controlled, a roll-off occurs when the system detects an important increase in impedance and the generator automatically shuts off, indicating that the desired ablation has been obtained, aiming to ensure ablation of consistent size according to the manufacturer's instructions for use: an ellipsoid of long axis $A_{manu} = 45$ mm and short axis $B_{manu} = 42$ mm.

Post-operative imaging: When the ablation is complete, the RFA electrode was removed, a multiphase contrast-enhanced CT scan was performed to evaluate the ablation zone and/or potential treatment-related complications.

Clinical Annotations. An experienced IR (more than 3 years of experience) manually annotated the needle position and segmented the tumor and the ablation zone on the intraoperative multiphasic enhanced CT-scans using ITK-SNAP [24], as illustrated in Fig. 1. To segment the tumor, the pre-treatment acquisition in which it was best visible was used, generally the arterial phase. The tumor was then segmented cut by cut in the 3 planes of space. The ablation zone was segmented the same way by using the post-treatment venous acquisition. Regarding the needle position, annotations were reported on the pre-treatment acquisition while observing the needle positioning venous CT scan acquired when the needle was at the final position. The two points correspond to the distal tip of the needle and the intersection between the needle and the skin. That method allows to accurately represent the needle direction compared to the tumor without requiring any registration steps.

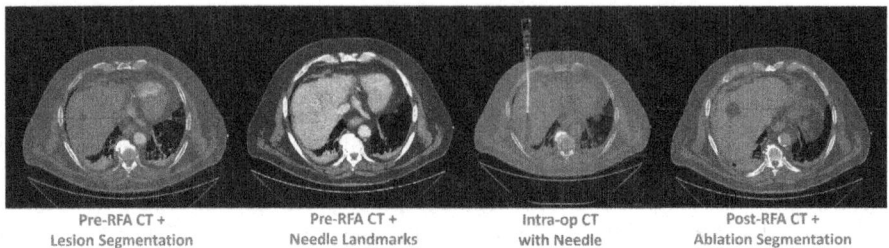

Fig. 1. Annotation Pipeline. The lesion is annotated on the pre-RFA CT, the needle landmarks (target point and a second point indicating the needle direction) are annotated on the pre-RFA CT while looking at the intra-op CT, and the ablation zone is segmented on the post-RFA CT. (Color figure online)

Patient Selection. To obtain a particularly high-quality dataset, we only selected patients that had not moved between the pre-operative and post-operative imaging. It is a valid assumption since patients were under general anesthesia and controlled breathing, and since the images were always acquired at the same breathing phase. This allowed to avoid any registration step, a major source of error as reported in several clinical evaluations [16–23, 25]. To this end, we applied a deep learning image-to-image network derived from [26] to automatically segment the livers from the pre-operative image and from the post-operative image. We computed the Dice Similarity Coefficient (DSC) and Hausdorff distance (HD) between the two liver segmentations obtained and discarded the cases where the DSC was below 89% or the HD was above 19mm as the quality of the segmentation or image alignment was deemed insufficient. Our dataset included patients who had several tumors and/or had previously received an ablation treatment for another tumor leading to selection errors when there was a mismatch between the tumor segmented on the pre-operative image and ablation zone segmented on the post-operative image. To only keep relevant cases, we computed the coverage percentage between the tumor and the ablation zone and discarded the cases where the coverage percentage (CP) was below 90% as we estimated that those cases presented an ablation segmentation not corresponding to the segmented tumor. Those exclusion criteria (DSC < 89% and HD > 19 mm between the liver segmentations and CP < 90% between the tumor and the ablation zone) were empirically determined based on a visual inspection of the images.

2.2 Fully Automatic Patient-Specific Thermal Ablation Modeling

A biophysics-based model is used to simulate the heat transfer in hepatic tissue while considering the heat sink effect from the blood vessels relying on patient-specific geometries. The model relies on 3 different parts:

Model of Patient Liver Anatomy. We automatically segment the liver parenchyma, portal veins, and hepatic veins using a deep learning image-to-image network derived from [26]. The resulting segmentations are illustrated in Fig. 2.

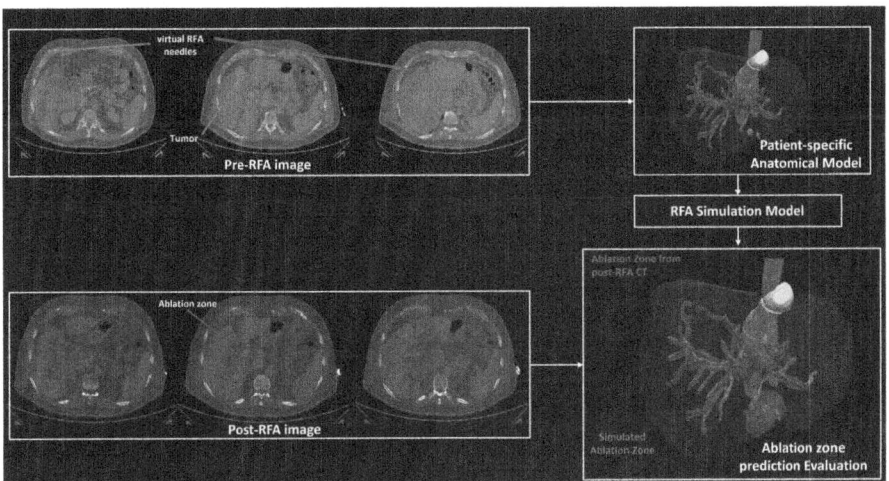

Fig. 2. Processing Pipeline. The patient-specific computational domain is created by running the automatic segmentation algorithm on the pre-RFA CT image. The RFA simulation is run in this domain, and the simulated ablation (**green**) is compared to the ground truth ablation zone (**red**) obtained from post-RFA CT images. (Color figure online)

Model of Heat Transfer in Hepatic Tissue. The bioheat equation Eq. (1) describes how the temperature propagates from the RF electrode through the liver, and considers the blood flow cooling effect of the major vessels [27]:

$$\rho c \partial T / \partial t = Q + \nabla \cdot (d \nabla T) + I_V \omega (T_{\text{blood}} - T) \quad (1)$$

In this equation, ρ, c, and d stand for density, specific heat capacity, and thermal conductivity of tissues, respectively; t is the time, T the tissue temperature, and Q the heat source term. In addition, the model accounts for the heat sink effect of the large hepatic vessels with a reaction term that uses the convection heat transfer coefficient ω and the blood temperature T_{blood} (assumed constant at 37 °C), as well as the vessel indicator function I_V. Finally, a Neumann boundary condition is applied at the border of the liver.

Model of Necrosis. Tissue necrosis is estimated based on the simulated temperature using a three-state model [28], which computes the variation of concentration of alive, vulnerable and dead cells over time. This model is fully coupled to the bioheat model since the heat capacity is updated based on the cell state as in [16].

Model of Heat Source. The source term Q was modeled by a Dirichlet boundary condition in an ellipsoid of fixed axis A_{source} and B_{source} mm centered at the RF electrode tip with a distal throw D_{source}, at temperature $T_{\text{source}} = 105$ °C for a duration of t_{abl} seconds followed by $t_{cooling}$ seconds without the Dirichlet boundary condition to reach a steady state.

Finally, to ensure a fair comparison with the manufacturer's estimation, we calibrated A_{source} and B_{source} ensuring that the model gave the exact same ablation zone when one single electrode was placed at the center of the computational domain without any blood vessels. In this case, the ablation zone was an ellipsoid of semi-axis A = 45 mm and B = 42 mm, corresponding to the ablation zone estimated by the manufacturer.

Implementation. Equation 1 was solved with a timestep of 0.1 s and a spatial resolution of 1 mm using a Lattice Boltzmann Method implementation for fast computation on GPU [16]. For all patients, the same tissue parameters were employed. Clinically relevant parameter values were chosen based on measurements from clinical studies [20] or according to manufacturer data, whereas technical parameters come from [16].

2.3 Computational Model Evaluation

The evaluation of the model predictions was done by comparing the simulated ablation zone to the patient-specific ground truth, as illustrated in Fig. 2 (right).

Evaluation Metrics. The DSC and HD between the predicted ablation zone and the ground truth were computed. The zones were also compared in terms of the Relative Volume Difference $RVD = \frac{|Vs - Vgt|}{Vgt}$ where Vs is the simulated ablation zone volume and Vgt, the ground truth volume.

Statistical Analysis. If normally distributed, result parameters were expressed as mean ± STD (95% CI: LL, UL) with STD, CI, LL and UL standing for Standard Deviation, Confidence Interval, Lower Limit of the confidence interval and Upper limit of the confidence interval respectively.

If non-normally distributed, result parameters were expressed as median (IQ: Q1, Q3) with IQ, Q1 and Q3 standing for interquartile, the 25th percentile and 75th percentile respectively.

The normality of data distribution was assessed with the Kolmogorov-Smirnov test.

To compare parameters differences between two models, an unpaired Student's t-test or a Mann-Whitney U test was applied, depending on data distribution. P-values < 0.05 were considered statistically significant.

3 Results

3.1 Data Processing

After the patient selection process, we had a consolidated dataset of 26 patients. On average, the patient tumor volume and long axis were 0.96 ± 0.88 cm^3 (95% CI: 0.61, 1.32) and 13.25 ± 4.60 mm (95% CI: 11.39, 15.11), the ablation volume and long axis were 23.23 ± 8.39 cm^3 (95% CI: 19.84, 26.61), and 46,31 ± 6.31 mm (95% CI: 43.76, 48.86) respectively with an ablation time ranging from 4 to 18 min. For 3 patients, the ablation time was unknown, and we assumed an ablation time of 12 min as prescribed by the manufacturer.

3.2 Repeatability Study

To determine the intra-operator variability, we repeated the exact same segmentation protocol under the same conditions and using the same method as described above on five random cases already performed three months earlier. The manual segmentation variability is reported in Table 1 in terms of volume and long axis errors, DSC and HD for the tumor and ablation zone and in terms of length and angle errors and distance for the needle.

Table 1. Repeatability study results in segmentation and needle annotations.

Difference in		Mean +/- std	Range: min – max
Pre-op tumor segmentation	Volume	29.5 +/- 22.2 %	4.9 – 54.1
	A (longest axis)	11.2 +/- 9.7 %	0.3 – 19.7
	DSC	79.0 +/- 9.9 %	64.8 – 87.3
	HD	4.2 +/- 2.1 mm	2.2 – 7.3
Post-op ablation segmentation	Volume	8.4 +/- 7.8 %	1.5 – 21.3
	A (longest axis)	6.1 +/- 4.3 %	1.4 – 10.6
	DSC	86.1 +/- 5.6 %	77.6 – 92.4
	HD	7.8 +/- 2.5 mm	4.7 – 11.0
Needle Annotation	Length	15.73 +/- 26.53 %	0.60 – 63.03
	Angle in 3D with [1,0,0]	6.87 +/- 5.57 %	0.61 – 14.01
	Angle with axial plane	27.74 +/- 37.28 %	0 – 100
	Angle within axial plane	6.27 +/- 4.88 %	0.61 – 11.39
	Target Point Distance	5.19 +/- 3.29 mm	1.18 – 8.63
	Entry Point Distance	12.73 +/- 11.11 mm	5.66 – 32.44

3.3 Distal Throw Optimization

As shown in Table 1, the needle annotations exhibited a large variability. For this reason, we decided to only rely on the direction formed by the two annotated points and to infer the actual target point from the data since the needle target point has a strong influence on the simulation results. To do so, we optimized D_{source} along the annotated needle direction for each patient to maximize the DSC between the manufacturer's estimation and the segmented post-op ablation. The mean DSC and HD improved from $44.86 \pm 12.89\%$ (95% CI: 39.66, 50.06) and 19.20 ± 4.22 mm (95% CI:17.49, 20.9) respectively when the annotated target point was used to $48.05 \pm 7.91\%$ (95% CI: 44.85, 51.25) and 17.48 ± 3.52 mm (95% CI: 16.06, 18.90) when the optimized distal throw per patient was used.

3.4 Simulated Ablation Zone Evaluation

Across the entire dataset, the DT model achieves a median RVD of 0.37 (IQ: 0.21, 0.51), representing a significant improvement over the manufacturer's estimate with a median RVD of 1.12 (IQ: 0.68–1.47). Although to a lesser extent, improvements are also observed in terms of DSC and HDD. The mean DSC increases from 48.05% ± 7.91 (95% CI: 44.85, 51.25) with the manufacturer's estimation to 50.49% ± 9.55 (95% CI: 46.63, 54.34) with the DT model. Similarly, the mean HDD increases slightly from 17.48 ± 3.52 mm (95% CI: 16.06, 18.90) to 17.82 ± 4.15 mm (95% CI: 16.14, 19.49). Figure 3 shows boxplots extending from Q1 to Q3, with a line at the median. It shows results for the full patient cohort (N = 26), as well as stratified results for patients with known ablation times (N = 23) and those for whom the standard ablation time was applied (N = 3).

Scatter plots of the digital twin (DT) predicted ablation volume and manufacturer's estimation against the ground truth post-op volume are shown in Fig. 4 (left). The volumes derived from the manufacturer's specifications remain constant in all cases while the DT predictions are closer to the ground truth. Figure 4 (right) illustrates how the ablation zone predicted by the DT model aligns well with the ground truth for patient 26.

The mean computational time to simulate 1 s of ablation was 0.43 ± 0.11 s (95% CI: 0.39, 0.47) on a NVIDIA RTX A4500 GPU with a mean memory footprint of 981.81 ± 259.66 MB (95% CI: 876.93, 1086.69).

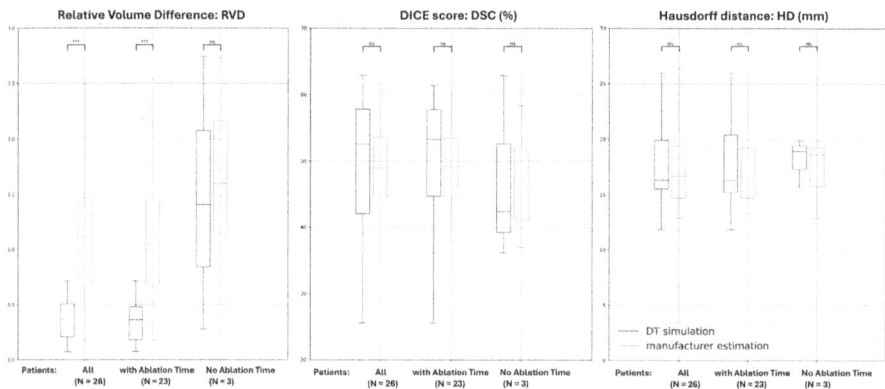

Fig. 3. Boxplots extending from Q1 to Q3, with a line at the median for the Relative Volume Difference, Dice Score and Hausdorff distance computed between the ground truth segmented ablation zone and the ablation zone estimated with the digital twin model **(green)** and with the manufacturer's estimation **(yellow).** Results are shown considering the entire dataset and separately for patients where the ablation time is known or unknown. Statistical Significance is shown with ns corresponding to a P-value ≥ 0.05 and *** to a P-value < 0.05. (Color figure online)

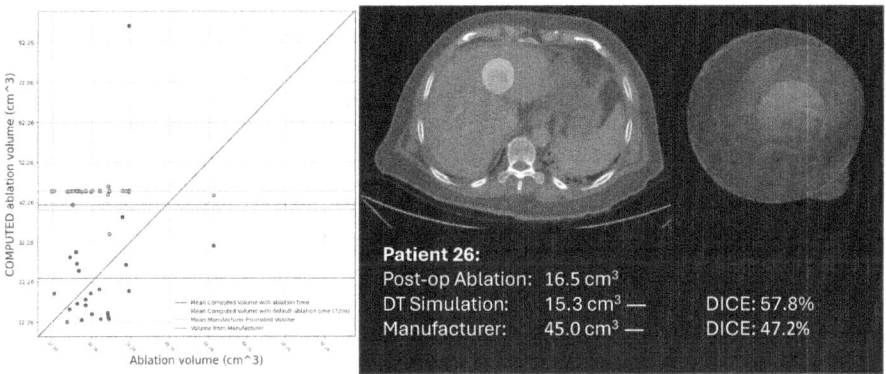

Fig. 4. (Left) Comparison between the ablation volume observed on the post-op CT and the computed ablation zone using the Digital twin model (**green**: patients with ablation time, **blue**: patients with standard ablation time) or using the estimation from the manufacturer (**yellow**). (Right) Patient 26 ablation zone predicted with the Digital twin model (**blue**) compares qualitatively well with ground truth (**red**) and better than the manufacturer's estimation (**yellow**). (Color figure online)

4 Discussion and Conclusion

The proposed digital twin model simulates RFA using patient-specific anatomical and needle information extracted from intra-operative CT scans acquired during regular clinical CT-guided ablation workflows. The fully automated pipeline enables scalable and reliable evaluation by removing manual steps that are prone to human error. It also allows direct analysis on intra-operative data, rather than relying on diagnostic images, eliminating the need for registration steps and thus facilitating a smoother integration into clinical workflows. The simulated ablation zones were compared against the ablation zone segmented on the post-ablation CT images. Improved results were achieved using the DT model, which accounts for the heat sink effect, in contrast to the manufacturer's estimation that overlooks this factor. Additionally, prescribing a personalized ablation time yielded better outcomes compared to using a standard duration, emphasizing the importance of collecting and integrating all available patient-specific information.

The variability observed in the repeatability study highlights the inherent difficulty to retrospectively segment the ablation zone on the post-RFA images acquired right after the intervention. This finding supports the assertion that achieving a DSC above 70% and a HD below 10 mm constitutes a reasonable benchmark for matching the segmentation accuracy of a trained radiologist. The repeatability study also revealed substantial variability in needle annotations, which prevented us from relying solely on the annotated target point. As a result, we optimized the distal throw parameter using the post-operative image. This represents a clear limitation of the current single-center retrospective study, which could be mitigated in future prospective studies by using an intra-operative tool to accurately mark the needle position.

In this study, we relied on nominal tissue parameters that do not account for the hepatic tissue heterogeneity, and hepatic perfusion was not incorporated into the model. We acknowledge these limitations and will aim to address them in future work to enhance physiological accuracy. Another limitation of the study is that we could not model the roll-offs since this information is not recorded.

By standardizing the planning process, liver digital twins can reduce the reliance on the operator experience at choosing the number of needles, placing them accurately and selecting the appropriate ablation settings (i.e., ablation power and duration) [5], thereby improving procedure consistency and democratizing the intervention success.

Disclosure of Interests. The authors have no competing interests to declare that are relevant to the content of this article to be updated if accepted.

Disclaimer. The concepts and information presented in this paper are based on research results that are not commercially available. Future commercial availability cannot be guaranteed.

References

1. Singal, A.G., et al.: AASLD practice guidance on prevention, diagnosis, and treatment of hepatocellular carcinoma. Hepatology **78**(6), 1922–1965 (2023)
2. Kok, H.P., et al.: Heating technology for malignant tumors: a review. Int. J. Hyperthermia **37**(1), 711–741 (2020)
3. Brace, C.L.: Radiofrequency and microwave ablation of the liver, lung, kidney, and bone: what are the differences? Curr. Probl. Diagn. Radiol. **38**(3), 135–143 (2009)
4. Laimer, G., et al.: Minimal ablative margin (MAM) assessment with image fusion: an independent predictor for local tumor progression in hepatocellular carcinoma after stereotactic radiofrequency ablation. Euro. Radiol. **30**, 2463–2472 (2020)
5. Crocetti, L., de Baére, T., Pereira, P.L., Tarantino, F.P.: CIRSE standards of practice on thermal ablation of liver tumours. Cardiovasc. Intervent. Radiol. **43**(7), 951–962 (2020)
6. Shady, W., et al.: Percutaneous radiofrequency ablation of colorectal cancer liver metastases: factors affecting outcomes-a 10-year experience at a single center. Radiology **278**(2), 601–611 (2016)
7. Liu, Z., Ahmed, M., Weinstein, Y., Yi, M., Mahajan, R.L., Goldberg, S.: Characterization of the RF ablation-induced 'oven effect': the importance of background tissue thermal conductivity on tissue heating. Int. J. Hyperthermia **22**(4), 327–342 (2006)
8. Livraghi, T., Goldberg, S.N., Lazzaroni, S., Meloni, F., Solbiati, L., Gazelle, G.S.: Small hepatocellular carcinoma: treatment with radio-frequency ablation versus ethanol injection. Radiology **210**(3), 655–661 (1999)
9. Nakazawa, T., et al.: Radiofrequency ablation of hepatocellular carcinoma: correlation between local tumor progression after ablation and ablative margin. Am. J. Roentgenol. **188**(2), 480–488 (2007)
10. Fang, Z., et al.: Radiofrequency ablation for liver tumors abutting complex blood vessel structures: treatment protocol optimization using response surface method and computer modeling. Int. J. Hyperth. **39**(1), 733–742 (2022)
11. Wright, A.S., Sampson, L.A., Warner, T.F., Mahvi, D.M., Lee, F.T., Jr.: Radiofrequency versus microwave ablation in a hepatic porcine model. Radiology **236**(1), 132–139 (2005)
12. Lu, D.S., Raman, S.S., Vodopich, D.J., Wang, M., Sayre, J., Lassman, C.: Effect of vessel size on creation of hepatic radiofrequency lesions in pigs: assessment of the heat sink effect. Am. J. Roentgenol. **178**(1), 47–51 (2002)
13. Lu, D.S., et al.: Influence of large peritumoral vessels on outcome of radiofrequency ablation of liver tumors. J. Vasc. Interv. Radiol. **14**(10), 1267–1274 (2003)

14. Kei, S.K., Rhim, H., Choi, D., Lee, W.J., Lim, H.K., Kim, Y.S.: Local tumor progression after radiofrequency ablation of liver tumors: analysis of morphologic pattern and site of recurrence. Am. J. Roentgenol. **190**(6), 1544–1551 (2008)
15. Rieder, C., Kroeger, T., Schumann, C., Hahn, H.K.: GPU-based real-time approximation of the ablation zone for radiofrequency ablation. IEEE Trans. Vis. Comput. Graph. **17**(12), 1812–1821 (2011)
16. Audigier, C., et al.: Efficient lattice Boltzmann solver for patient-specific radiofrequency ablation of hepatic tumors. IEEE Trans. Med. Imag. **34**(7), 1576–1589 (2015)
17. Voglreiter, P., et al.: RFA guardian: comprehensive simulation of radiofrequency ablation treatment of liver tumors. Sci. Rep. **8**(1), 787 (2018)
18. Frackowiak, B., et al.: First validation of a model-based hepatic percutaneous microwave ablation planning on a clinical dataset. Sci. Rep. **13**(1), 16862 (2023)
19. Singh, S., Melnik, R.: Thermal ablation of biological tissues in disease treatment: a review of computational models and future directions. Electromagn. Biol. Med. **39**(2), 49–88 (2020)
20. Mariappan, P., et al.: GPU-based RFA simulation for minimally invasive cancer treatment of liver tumours. Int. j. Comput. Assist. Radiol. Surg. **12**, 59–68 (2017)
21. Moche, M., et al.: Clinical evaluation of in silico planning and real-time simulation of hepatic radiofrequency ablation (ClinicIMPPACT Trial). Euro. Radiol. **30**, 934–942 (2020)
22. Audigier, C., et al.: Comprehensive preclinical evaluation of a multi-physics model of liver tumor radiofrequency ablation. Int. J. Comput. Assist. Radiol. Surg. **12**, 1543–1559 (2017)
23. Audigier, C., Mohaiu, A.T., Alzaga, A., Bale, R., Mansi, T.: A comparative study on computational models of multi-electrode radiofrequency ablation of large liver tumors. Int. J. Comput. Assist. Radiol. Surg. **17**(8), 1489–1496 (2022)
24. Yushkevich, P.A., et al.: User-guided 3D active contour segmentation of anatomical structures: significantly improved efficiency and reliability. Neuroimage **31**(3), 1116–1128 (2006)
25. van Amerongen, M.J., et al.: Software-based planning of ultrasound and CT-guided percutaneous radiofrequency ablation in hepatic tumors. Int. J. Comput. Assist. Radiol. Surg. **16**(6), 1051–1057 (2021)
26. Yang, D., et al.: Automatic liver segmentation using an adversarial image-to-image network. In: International Conference on Medical Image Computing and Computer-Assisted Intervention. Springer, pp. 507–515 (2017)
27. Pennes, H.H.: Analysis of tissue and arterial blood temperatures in the resting human forearm. J. Appl. Physiol. **1**(2), 93–122 (1948)
28. ONeill, D., et al.: A three-state mathematical model of hyperthermic cell death. Ann. Biomed. Eng. **39**, 570–579 (2011)

Acoustic Simulation with Deep Learning for Low-Intensity Transcranial Focused Ultrasound Digital Twins

Minjee Seo[1], Minwoo Shin[2], Gunwoo Noh[3], Seung-Schik Yoo[4], and Kyungho Yoon[1,5](✉)

[1] School of Mathematics and Computing (Computational Science and Engineering), Yonsei University, Seoul, Republic of Korea
yoonkh@yonsei.ac.kr
[2] Department of Software, Yonsei University, Wonju, Republic of Korea
[3] School of Mechanical Engineering, Korea University, Seoul, Republic of Korea
[4] Department of Radiology, Brigham and Women's Hospital, Harvard Medical School, Boston, MA, USA
[5] Graduate School of Artificial Intelligence, Pohang University of Science and Technology (POSTECH), Pohang, Republic of Korea

Abstract. Transcranial focused ultrasound (tFUS) is a promising therapeutic modality capable of delivering concentrated acoustic energy to targeted brain regions. A major challenge lies in the significant distortion of the ultrasound beam caused by the skull, leading to an unpredictable shift in location and intensity of the acoustic focus. For the treatment procedure to be both safe and effective, estimating the distorted acoustic focus in real-time is essential. However, existing acoustic simulation methods to predict the acoustic field are computationally too intensive for real-time clinical use. To address this gap, we propose a deep learning-based real-time acoustic simulation method to establish a low-intensity focused ultrasound (LIFU) digital twin. Our approach rapidly estimates intracranial acoustic pressure fields during treatment by taking the acoustic free-field, skull image, and transducer placement as input using multi-modal neural networks. We evaluated model performance on both seen and unseen skull anatomies to verify generalizability. Our models achieved inference times of approximately 23 milliseconds, confirming their suitability for real-time simulation. Our method enables the construction of a digital twin framework that dynamically reflects the ongoing therapeutic state, providing a foundation for data-driven, adaptive LIFU treatment strategies. The code is available at: https://github.com/CMME-Lab/LIFUSimul-DL.git.

Keywords: Transcranial focused ultrasound · Acoustic simulation · Deep learning

1 Introduction

Transcranial focused ultrasound (tFUS) has emerged as a promising non-invasive neuromodulation technique that enables the precise delivery of acoustic energy to deep brain targets [13,16]. Its ability to focus ultrasound through the skull without surgical intervention has led to growing interest in both therapeutic and research contexts [26]. In particular, low-intensity focused ultrasound (LIFU) has recently gained considerable attention as a novel approach for treating various conditions through non-invasive brain stimulation [1,8].

One of the major obstacles in tFUS is the precise targeting of the acoustic focus, which becomes difficult due to the complex interaction between ultrasound waves and the skull [17]. As ultrasound propagates, its behavior is heavily influenced by the acoustic properties of the intervening tissues. Given the highly heterogeneous composition of the skull, the transmitted waves undergo significant distortion and energy loss [9,10]. This variability makes it particularly challenging to pinpoint the actual focal area within the brain during tFUS procedures [4,25]. Therefore, monitoring the intracranial acoustic field during treatment procedure is critical for ensuring both patient safety and therapeutic efficacy.

Digital twin is a computational system designed to replicate the properties and dynamic behavior of real-world physical entities within a simulated environment [6]. This technology has garnered significant attention for its potential applications in tailoring medical treatments to individual patients and supporting precision healthcare delivery [7,11]. In the context of LIFU therapy, digital twins are particularly well-suited for predicting intracranial acoustic pressure fields that vary depending on patient-specific skull anatomy and transducer placement. To enable such predictions, numerical simulation methods offer a viable solution, as they can accurately simulate acoustic wave propagation by incorporating complex anatomical structures and acoustic properties of the skull into wave equation solvers [20,25]. However, their substantial computational burden poses a significant barrier to routine real-time clinical application [12]. As a result, there is increasing interest in developing more efficient simulation frameworks that offer real-time feedback without compromising on accuracy [20].

Motivated by these demands, we propose a deep learning-based acoustic simulation method for a data-driven LIFU digital twin framework. Our method aims to rapidly predict intracranial acoustic fields by integrating multi-modal inputs, including acoustic free-fields, patient-specific medical images of the skull, and transducer placement information. The network was trained on data collected from 11 human subjects, where acoustic fields were computed by acoustic simulations using the k-space method based on the skull CT volumes of each subject. Performance was evaluated on both foreseen data from 8 subjects included in training and unseen data from the remaining 3 subjects. Results demonstrate the feasibility of real-time, patient-specific simulation as a core component of digital twin systems for LIFU therapy.

2 Methods

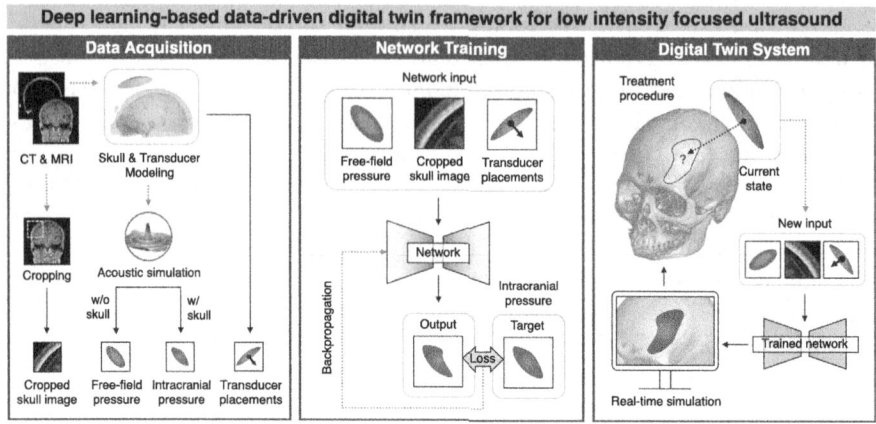

Fig. 1. Overview of our digital twin framework.

2.1 Data Processing

Skull CT and MR. Three-dimensional computed tomography (CT) and magnetic resonance (MR) images of the skull were acquired from 11 human subjects. To enable precise co-registration between modalities, four fiducial markers were affixed to each subject's scalp. CT volumes were acquired at an isometric resolution of $0.5 \times 0.5 \times 0.5$ mm^3, and T1-weighted MR images were captured using a 3D GRAPPA sequence with a voxel size of $1.0 \times 1.0 \times 1.0$ mm^3, and subsequently resampled to $0.5 \times 0.5 \times 0.5$ mm^3 using Lanczos interpolation to ensure spatial consistency with the CT volumes. Co-registration between CT and MR images was performed via a point-based rigid-body transformation algorithm [15] (Fig. 1).

Modeling Anatomical Geometries. CT images were thresholded based on the Hounsfield units (ϕ_{ijk}) to extract the anatomical structure of skull. Voxels with $\phi_{ijk} \leq 0$ were classified as water, those with $0 < \phi_{ijk} \leq 1000$ as trabecular bone, and voxels with $1000 \leq \phi_{ijk}$ as cortical bone. The acoustic properties in each region—namely, the speed of sound (c_{ijk}), density (ρ_{ijk}), and attenuation coefficient (a_{ijk})—were assigned based on values reported in previous studies [3,18], as shown in Table 1. The thalamus region of the brain was extracted from the MR images and defined as the sonication target.

Modeling Transducer. A bowl-shaped single-element transducer operating at a frequency of 250 kHz was modeled. Its physical specifications were defined by a

Table 1. Acoustic properties assigned based on HU-based segmentation.

	water	trabecular bone	cortical bone
c_{ijk}	1500 m/s	2140 m/s	2384 m/s
ρ_{ijk}	1000 kg/m^3	$1000 + 1.19\phi_{ijk}$ kg/m^3	2190 kg/m^3
α_{ijk}	0 Np/m	33 Np/m	33 Np/m

diameter of 75 mm and both a radius of curvature and focal length of 83 mm. For each subject, 400 transducer placements were randomly generated to maintain a fixed distance of 83 mm from the target and alignment with the skull surface normal.

Acoustic Simulation. The governing equations describing wave propagation in an inhomogeneous and absorbing medium are given as follows:

$$\begin{aligned}
\frac{\partial \mathbf{u}}{\partial t} &= -\frac{1}{\rho_0} \nabla p, \\
\frac{\partial \rho}{\partial t} &= -\rho_0 \nabla \cdot \mathbf{u} - \mathbf{u} \cdot \nabla \rho_0, \\
p &= c_0^2 \left(\rho + \mathbf{d} \cdot \nabla \rho_0 - L\rho \right),
\end{aligned} \quad (1)$$

where $\mathbf{u}$ is the acoustic particle velocity, p is the acoustic pressure, ρ is the acoustic density, ρ_0 is the ambient density, c_0 is the isentropic sound speed, $\mathbf{d}$ is the acoustic particle displacement, and the operator L is a linear integro-differential operator that accounts for acoustic absorption and dispersion that follows a frequency power law [2]. To obtain the acoustic pressure field, simulations via k-space method [21] were performed using the k-Wave MATLAB toolbox [23]. The time step size for the computation was determined to satisfy the Courant-Friedrichs-Lewy (CFL) condition [21].

2.2 Dataset Construction

Acoustic Fields. In this study, the acoustic field is categorized into two components: the free-field, which serves as the network input, and the intracranial field, which is the network target. Intracranial fields were simulated using the skull structures and transducer placements defined in Sect. 2.1. The resulting volume was then cropped to a field of view (FOV) of $112 \times 112 \times 112$ voxels centered around the target region.

The acoustic free-field is defined as the acoustic pressure field generated during propagation in a homogeneous medium. This models the basic propagation pattern in the absence of skull-induced aberrations and serves as a baseline for predicting the corresponding intracranial acoustic field. Since the acoustic free-field is solely determined by the transducer's geometry and operating frequency, it remains constant for a given transducer. Therefore, we computed a reference free-field by simulating wave propagation in a homogeneous water medium, and then rotated it according to the transducer orientation vectors defined in Sect. 2.1 to align with each target intracranial acoustic field.

Skull Images. To enable the network to effectively model the skull-induced deformation of the acoustic field, medical imaging data representing the skull geometry were incorporated as part of the network input. The CT and MR images, as described in Sect. 2.1, were each cropped to match the FOV of the acoustic field, centered at the point where the ultrasound beam intersects the skull surface.

Transducer Placements. To explicitly provide the spatial configuration of transducer to the network, we defined the transducer placement input vector. This was generated by concatenating the position vector $[T_x, T_y, T_z]$, representing the transducer's center location, and the orientation vector $[\theta_x, \theta_y, \theta_z]$, indicating the direction toward the target region.

2.3 Deep Learning Network Architectures

Multi-modal Feature Processing Modules. To extract and effectively integrate information from distinct input modalities, we designed a multi-modal feature processing module. For transducer placement, a Fourier feature embedding layer [22] followed by a linear layer was used to effectively project low-dimensional coordinate information into a higher-dimensional space. For both the free-field and skull image data, feature extraction was performed using two types of encoding blocks: one based on Convolutional Neural Networks (CNNs) [24] and another based on Swin Transformers [14]. This comparative approach was designed to determine whether the CNN's strength in capturing local features or the Swin Transformer's capacity for modeling long-range global dependencies would yield more effective feature representations for the given modalities. After passing through four encoding blocks, each modality was transformed to the same spatial dimensions as the transducer placement. The features were then processed with self-attention and subsequently integrated, followed by further processing through additional self-attention and two encoding blocks.

Decoding Modules. Each decoding block consists of trilinear interpolation followed by a convolution, aiming to reconstruct the original spatial resolution from the integrated features. We explored two primary network architectures for reconstruction: an Autoencoder (AE) [5] structure, which relies solely on the decoding modules for reconstruction, and a U-Net [19] structure, which incorporates skip connections to integrate features from the encoding stage. Specifically, the AE was tested with the CNN-based encoding block, while the U-Net architecture was implemented using CNN-based encoding blocks, while the U-Net architecture was evaluated with both CNN- and Swin Transformer-based encoders, resulting in a total of three distinct network architectures. A detailed overview of the network is illustrated in Fig. 2.

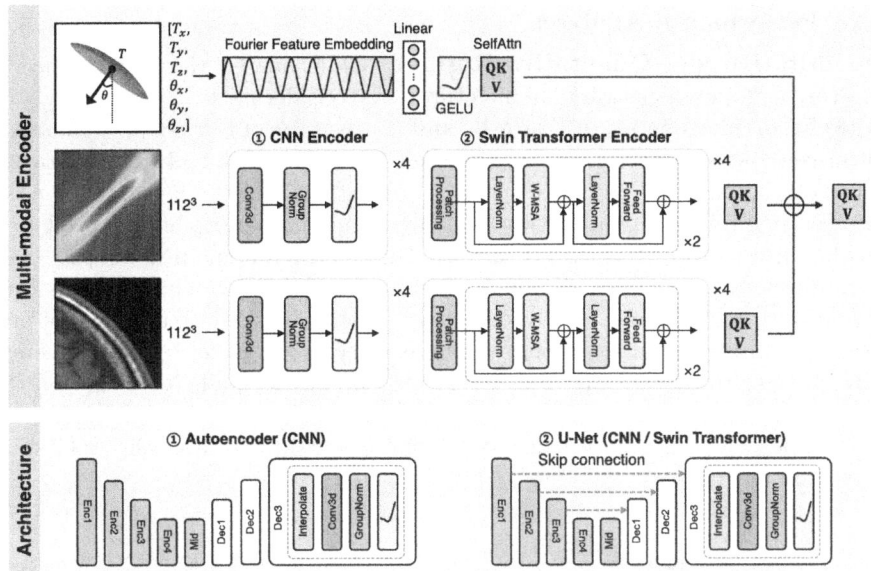

Fig. 2. Illustration of multi-modal feature processing modules and networks.

3 Results

3.1 Experimental Settings

Dataset Splitting. Due to considerable anatomical variability, evaluating generalization to unseen skulls is essential. To assess this, we allocated data from 8 of the 11 subjects to the training and foreseen test sets by splitting each subject's data in a 8:2 ratio. The remaining 3 subjects' data were designated as the unseen test set.

Optimization Settings. Each network was trained by optimizing the mean squared error (MSE) between the predicted and target acoustic fields using the Adam optimizer. Training was conducted over 200 epochs, with an initial learning rate of 0.0002 for CNN-based models and 0.0005 for the Swin Transformer model. A learning rate scheduling strategy was applied, linearly decaying the learning rate based on the current epoch after the first 100 epochs.

Evaluation Metrics. In this study, we employed two evaluation metrics to assess model performance: (1) **Dice** score to quantify the overlap between focal regions of predicted and target acoustic fields, and (2) mean absolute percentage error $\mathbf{E_p}$ of peak pressure amplitude.

3.2 Performance Analyses

Quantitative and Qualitative Results. We evaluated the performance of the three proposed network architectures by considering two factors: (1) the type of skull image input (CT or MR) and (2) the type of test set used (foreseen or unseen). As summarized in Table 2, the CNN U-Net architecture consistently achieved the highest Dice score across all conditions and the lowest $\mathbf{E_p}$ for the foreseen test set. While all networks showed minimal performance degradation on the unseen test set when CT images were used as input, a noticeable drop in performance was observed when MR images were used. This suggests that CT images, which more clearly delineate the crucial skull structures for predicting skull-induced aberrations, are more effective for model training, whereas MR images primarily depict soft tissues excluding the skull. Nevertheless, the results also indicate that simulations based solely on MR inputs remain feasible with some degree of performance compromise. Qualitative analysis results are presented in Fig. 3.

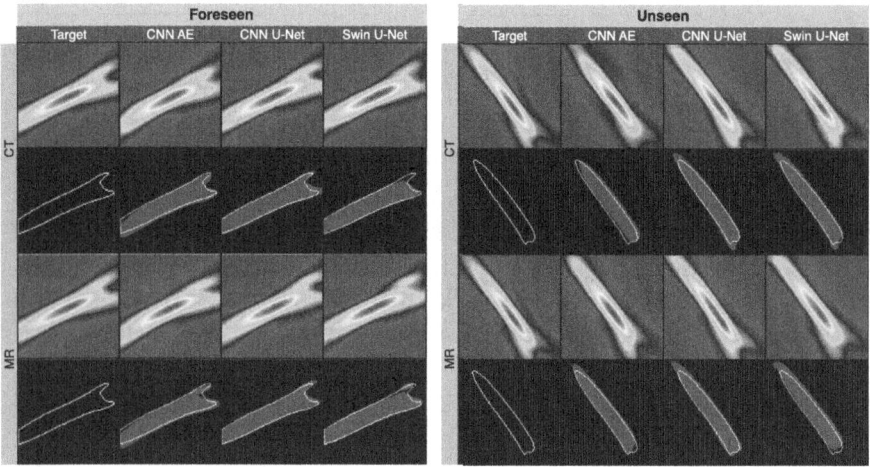

Fig. 3. Illustration of predicted and target intracranial fields, along with their corresponding focal regions. In the focal region, white edge indicates the target, while solid overlay indicates the predicted region.

Network Efficiency Analysis. To validate the efficiency of each network, we measured both inference time and the number of trainable parameters. As summarized in Table 3, CNN-based networks offer highly accurate simulation results with a small number of parameters and exceptionally fast inference times. While the Swin Transformer U-Net architecture has a considerably larger number of parameters and slower inference times due to its inherent design, it still achieves

Table 2. Comparison of performance using foreseen and unseen test data.

Modality	Network	Foreseen		Unseen	
		Dice (%) ↑	$\mathbf{E_p}$ (%) ↓	Dice (%) ↑	$\mathbf{E_p}$ (%) ↓
CT	CNN AE	87.49 ± 3.30	1.69 ± 1.13	86.00 ± 3.48	**4.44 ± 2.11**
	CNN U-Net	**93.73 ± 2.40**	**0.71 ± 0.59**	**88.64 ± 3.71**	5.29 ± 2.41
	Swin U-Net	92.10 ± 3.39	1.28 ± 1.14	84.33 ± 4.63	6.66 ± 3.11
MR	CNN AE	89.57 ± 3.33	1.87 ± 1.05	78.01 ± 7.68	5.99 ± 3.51
	CNN U-Net	**93.09 ± 2.95**	**0.75 ± 0.67**	**79.48 ± 7.12**	6.02 ± 3.55
	Swin U-Net	91.26 ± 3.70	1.25 ± 1.18	75.71 ± 8.04	**4.83 ± 3.29**

approximately 13,000 times faster inference than conventional simulation methods. This demonstrates its feasibility for real-time inference, which is crucial for building a digital twin framework.

Table 3. Comparison of number of parameters and simulation time.

Methods	Parameters (M)	Simulation Time (s)
k-Wave	–	308.033
CNN AE	11.6	0.00227
CNN U-Net	11.6	0.00230
Swin U-Net	38.8	0.0232

Ablation Study. We conducted an ablation study to determine the impact of each input data modality on the network's performance. Evaluating the foreseen

Table 4. Comparison of performance using different input data.

Inputs	Network	Dice (%) ↑	$\mathbf{E_p}$ (%) ↓
Free-Field	CNN AE	78.60 ± 10.18	4.10 ± 3.34
	CNN U-Net	80.46 ± 10.63	3.23 ± 2.83
	Swin U-Net	83.42 ± 6.66	2.88 ± 2.31
Free-Field + CT	CNN AE	87.33 ± 4.05	2.73 ± 1.48
	CNN U-Net	94.79 ± 2.17	0.63 ± 0.55
	Swin U-Net	93.13 ± 3.52	1.39 ± 1.15
Free-Field + CT + Transducer	CNN AE	87.49 ± 3.30	1.69 ± 1.13
	CNN U-Net	93.73 ± 2.40	0.71 ± 0.59
	Swin U-Net	92.10 ± 3.39	1.28 ± 1.14

test set with using CT as an input skull image, we observed a significant improvement in prediction accuracy when skull image data was included, compared to using only the free-field, as shown in Table 4. In contrast, the addition of transducer placement information did not lead to a significant performance enhancement. This suggests that the transducer placement information may already be implicitly contained in the free-field and skull image inputs. However, we anticipate that providing such information will be beneficial when exploring a wider range of transducer geometries and frequencies.

4 Conclusion

In this study, we propose a deep learning-based real-time acoustic simulation method for implementing a digital twin framework in LIFU therapy. By integrating acoustic free-field, skull image, and transducer placement as network inputs, our networks effectively generate intracranial acoustic fields that account for skull-induced aberrations. Our approach enables significantly faster simulation compared to conventional methods, achieving over 93% prediction accuracy, laying the groundwork for future real-time digital twin systems. By integrating an optical tracking system, the transducer's position can be overlaid onto medical images in real-time, simultaneously visualizing the predicted intracranial acoustic field for intuitive and interactive digital twin realization during LIFU procedures. Future research includes the expansion of the framework by considering varying transducer geometries and frequencies and by developing a generalized network capable of effectively handling such diversity.

Acknowledgments. The work was supported by the National Research Foundation of Korea (NRF) funded by the Korean government (MSIT) under Grants RS-2024-00335185 and RS-2023-00220762.

Disclosure of Interests. The authors declare that they have no known competing financial interests or personal relationships that could have appeared to influence the work reported in this paper.

References

1. Blackmore, J., Shrivastava, S., Sallet, J., Butler, C.R., Cleveland, R.O.: Ultrasound neuromodulation: a review of results, mechanisms and safety. Ultrasound Med. Biol. **45**(7), 1509–1536 (2019). https://doi.org/10.1016/j.ultrasmedbio.2018.12.015
2. Chen, W., Holm, S.: Fractional Laplacian time-space models for linear and nonlinear lossy media exhibiting arbitrary frequency power-law dependency. J. Acoust. Soc. Am. **115**, 1424–30 (2004). https://doi.org/10.1121/1.1646399
3. Deffieux, T., Konofagou, E.E.: Numerical study of a simple transcranial focused ultrasound system applied to blood-brain barrier opening. IEEE Trans. Ultrason. Ferroelectr. Freq. Control **57**(12), 2637–2653 (2010). https://doi.org/10.1109/TUFFC.2010.1738

4. Ghanouni, P., et al.: Transcranial MRI-guided focused ultrasound: a review of the technologic and neurologic applications. AJR Am. J. Roentgenol. **205**(1), 150–159 (2015). https://doi.org/10.2214/AJR.14.13632. Jul
5. Hinton, G.E., Salakhutdinov, R.R.: Reducing the dimensionality of data with neural networks. Science **313**(5786), 504–507 (2006). https://doi.org/10.1126/science.1127647
6. Jones, D., Snider, C., Nassehi, A., Yon, J., Hicks, B.: Characterising the digital twin: a systematic literature review. CIRP Manuf. Sci. Tech. **29**, 36–52 (2020). https://doi.org/10.1016/j.cirpj.2020.02.002
7. Katsoulakis, E., et al.: Digital twins for health: a scoping review. NPJ Digit. Med. **7**(1), 77 (2024). https://doi.org/10.1038/s41746-024-01073-0. Mar
8. Keihani, A., et al.: Transcranial focused ultrasound neuromodulation in psychiatry: main characteristics, current evidence, and future directions. Brain Sci. **14**(11) (2024). https://doi.org/10.3390/brainsci14111095
9. Kyriakou, A., Neufeld, E., Werner, B., Paulides, M.M., Szekely, G., Kuster, N.: A review of numerical and experimental compensation techniques for skull-induced phase aberrations in transcranial focused ultrasound. Int. J. Hyperthermia **30**(1), 36–46 (2014). https://doi.org/10.3109/02656736.2013.861519, pMID: 24325307
10. Kyriakou, A., Neufeld, E., Werner, B., Székely, G., Kuster, N.: Full-wave acoustic and thermal modeling of transcranial ultrasound propagation and investigation of skull-induced aberration correction techniques: a feasibility study. J. Ther. Ultrasound **3**(1), 11 (2015). https://doi.org/10.1186/s40349-015-0032-9. Jul
11. Laubenbacher, R., Mehrad, B., Shmulevich, I., Trayanova, N.: Digital twins in medicine. Nat. Comput. Sci. **4**(3), 184–191 (2024). https://doi.org/10.1038/s43588-024-00607-6. Mar
12. Leung, S.A., Webb, T.D., Bitton, R.R., Ghanouni, P., Butts Pauly, K.: A rapid beam simulation framework for transcranial focused ultrasound. Sci. Rep. **9**(1), 7965 (2019). https://doi.org/10.1038/s41598-019-43775-6. May
13. Lipsman, N., et al.: MR-guided focused ultrasound thalamotomy for essential tremor: a proof-of-concept study. Lancet Neurol. **12**(5), 462–468 (2013). https://doi.org/10.1016/S1474-4422(13)70048-6
14. Liu, Z., et al.: Swin Transformer: Hierarchical vision transformer using shifted windows (2021). https://doi.org/10.48550/arXiv.2103.14030
15. Moghari, M.H., Abolmaesumi, P.: Point-based rigid-body registration using an unscented Kalman filter. IEEE Trans. Med. Imaging **26**(12), 1708–1728 (2007). https://doi.org/10.1109/tmi.2007.901984
16. Naor, O., Krupa, S., Shoham, S.: Ultrasonic neuromodulation. J. Neural Eng. **13**, 031003 (2016). https://doi.org/10.1088/1741-2560/13/3/031003
17. Pasquinelli, C., Hanson, L.G., Siebner, H.R., Lee, H.J., Thielscher, A.: Safety of transcranial focused ultrasound stimulation: a systematic review of the state of knowledge from both human and animal studies. Brain Stimul. **12**(6), 1367–1380 (2019). https://doi.org/10.1016/j.brs.2019.07.024
18. Pichardo, S., Sin, V.W., Hynynen, K.: Multi-frequency characterization of the speed of sound and attenuation coefficient for longitudinal transmission of freshly excised human skulls. Phys. Med. Biol. **56**(1), 219 (2010). https://doi.org/10.1088/0031-9155/56/1/014
19. Ronneberger, O., Fischer, P., Brox, T.: U-Net: convolutional networks for biomedical image segmentation. CoRR abs/1505.04597 (2015). https://doi.org/10.48550/arXiv.1505.04597

20. Shin, M., Peng, Z., Kim, H.J., Yoo, S.S., Yoon, K.: Multivariable-incorporating super-resolution residual network for transcranial focused ultrasound simulation. Comput. Methods Programs Biomed. **237**, 107591 (2023). https://doi.org/10.1016/j.cmpb.2023.107591
21. Tabei, M., Mast, T.D., Waag, R.: A k-space method for coupled first-order acoustic propagation equations. J. Acoust. Soc. Am. **111**, 53–63 (2002). https://doi.org/10.1121/1.1421344
22. Tancik, M., et al.: Fourier features let networks learn high frequency functions in low dimensional domains (2020). https://doi.org/10.48550/arXiv.2006.10739
23. Treeby, B.E., Cox, B.T.: k-Wave: MATLAB toolbox for the simulation and reconstruction of photoacoustic wave fields. J. Biomed. Opt. **15**(2), 021314 (2010). https://doi.org/10.1117/1.3360308
24. Yamashita, R., Nishio, M., Do, R.K.G., Togashi, K.: Convolutional neural networks: an overview and application in radiology. Insights Imaging **9**(4), 611–629 (2018). https://doi.org/10.1007/s13244-018-0639-9. Aug
25. Yoon, K., Lee, W., Croce, P., Cammalleri, A., Yoo, S.S.: Multi-resolution simulation of focused ultrasound propagation through ovine skull from a single-element transducer. Phys. Med. Biol. **63**, 105001 (2018). https://doi.org/10.1088/1361-6560/aabe37
26. Zhang, T., Pan, N., Wang, Y., Liu, C., Hu, S.: Transcranial focused ultrasound neuromodulation: a review of the excitatory and inhibitory effects on brain activity in human and animals. Front. Hum. Neurosci. **Volume 15 - 2021** (2021). https://doi.org/10.3389/fnhum.2021.749162

Towards Digital Twin of RF Ablation: Real-Time Prediction of Time-Dependent Thermal Effects Using Transformer

Seonaeng Cho[1], Minjee Seo[1], Minwoo Shin[2], and Kyungho Yoon[1(✉)]

[1] School of Mathematics and Computing (Computational Science and Engineering), Yonsei University, Seoul, Republic of Korea
yoonkh@yonsei.ac.kr

[2] Division of Software, Yonsei University, Wonju, Republic of Korea

Abstract. Radiofrequency ablation (RFA) has emerged as a promising minimally invasive technique for tumor treatment, offering reduced recovery time and postoperative pain. Accordingly, accurately simulating ablation dynamics—including electrode positioning and timing—is critical for minimizing recurrence and preventing damage to surrounding healthy tissue. While a numerical model provides physically reliable predictions, it is inherently slow and computationally intensive, rendering it unsuitable for real-time clinical applications. To address this issue, we propose a transformer-based U-Net (UNETR) model trained on numerically computed thermal effects to enable real-time prediction of the spatiotemporal dynamics of 3D ablation regions and temperature maps. The model is trained on data generated from a multi-physics simulation incorporating electrostatic field analysis, bio-heat transfer, and cell necrosis modeling, based on breast cancer MR images. It achieved a root mean square error (RMSE) of 0.7160 for temperature distribution and a Dice score of 94.38% for ablation regions on previously unseen MR anatomies, demonstrating strong generalization across anatomical variations. Furthermore, inference time was drastically reduced from 76.23 s, required by conventional numerical methods, to just 0.047 s, enabling real-time performance. These results demonstrate the feasibility of a digital twin for RFA, which holds promise for improving the safety and efficacy of personalized therapy. The code is available at: https://github.com/SeonAengCho/RFA-simulation-model.git.

Keywords: Radiofrequency ablation · Thermal effect simulation · Deep learning

1 Introduction

Radiofrequency ablation (RFA) is a minimally invasive procedure in which a high-frequency electrical current delivered via an inserted electrode induces thermal necrosis of tumor tissue [11,22]. Compared to conventional surgical resection,

RFA requires only minimal skin incision, resulting in reduced bleeding, less postoperative pain, and shorter recovery. It is applicable to anatomically challenging sites and suitable for patients ineligible for surgery. Due to limited visualization of the target during the procedure, careful pre-procedural planning—including power settings, electrode placement, and heating duration—is essential to ensure complete ablation while minimizing damage to healthy tissue [9]. However, current RFA procedures often rely on generalized elliptical damage models provided by device manufacturers to predict the ablation zone. As a result, treatment planning does not adequately account for patient-specific anatomical and physiological variations, and clinical outcomes may vary depending on the operator's experience [7].

Digital twin technology replicates real-world systems in a virtual environment and is used in medicine for patient-specific diagnosis and treatment [21,24]. In RFA, this provides real-time feedback on thermal effects, addressing the aforementioned limitations. Achieving this requires real-time simulation of the procedure's inherent multiphysics processes. While traditional models—integrating electrical conduction, bioheat transfer, and tissue damage—accurately predict temperature and ablation zones [23,27], they are computationally expensive, often taking seconds to minutes even with GPU acceleration, making them unsuitable for real-time clinical use, especially in high-resolution or time-dependent settings [10,17,25].

Recent advances combine deep learning with physics-based simulations to reduce the high computational cost of modeling complex processes [18,19]. These methods preserve the accuracy and interpretability of traditional models while enabling faster inference. In healthcare, where real-time, patient-specific predictions are vital, integrating medical imaging, biophysical modeling, and data-driven learning offers a powerful approach for rapid prediction of treatment outcomes.

In this study, we propose a deep learning-based simulation framework for real-time prediction of RFA. Traditional RFA simulations require complex multi-stage numerical computations—including electric potential calculation, thermal diffusion analysis, and tissue necrosis modeling. To replace this process, we design a neural network that accurately approximates the outputs of conventional simulations. The model takes the patient's tumor morphology and electrode position as input, and predicts both the temperature distribution and ablation zone derived from physics-based simulations. The output consists of 18 channels corresponding to sequential time steps, allowing the model to estimate the entire ablation process in a single forward pass. To effectively capture spatially localized but clinically important features—such as electrode position—we adopt a transformer-based U-Net architecture [5]. Two separate networks with this structure are trained independently: one for temperature prediction and the other for damage region estimation. Based on this approach, we present a high-accuracy, real-time inference model for RFA simulation with the following key advantages: (1) accurate approximation of conventional numerical results with low prediction error, (2) efficient prediction of thermal evolution across 18 time steps in a single forward pass, and (3) inference speed fast enough for potential integration into clinical workflows.

2 Method

This section presents a detailed explanation of the proposed method. As shown in Fig. 1, the overall pipeline consists of three sequential stages: data preparation, numerical simulation, and neural network training. The implementation details of each stage are described in the following subsections.

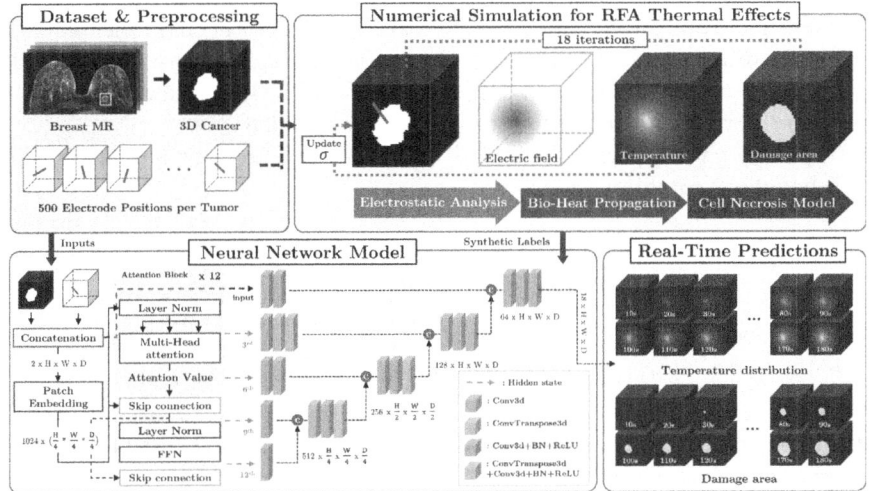

Fig. 1. Overview of the proposed real-time simulation of thermal effects in RFA

2.1 Dataset and Preprocessing

MR images from 13 breast cancer patients were obtained from a publicly available dataset [16], including DCE-MRI scans and binary tumor segmentation masks annotated by expert radiologists. All images were resampled to isotropic voxels of $1.0 \times 1.0 \times 1.0$ mm^3 using the Lanczos filter [2]. A $40 \times 40 \times 40$ mm^3 region centered at the tumor was extracted for each patient and defined as the simulation domain. To reduce overfitting and focus on the tumor region, segmentation masks were used instead of the original MR images. Within each domain, 500 electrode insertion points were generated per patient by randomly sampling spatial positions and directions. To ensure greater diversity in needle placement rather than tumor morphology, multiple needle positions were used for each tumor despite the limited number of tumors.

2.2 Numerical Simulation For Synthetic Labels

Electrostatic Analysis for Electric Potential. The simulation begins by placing an electrode within the 3D tumor geometry and computing the resulting electric potential. Given that RFA operates at frequencies (450–550 kHz) with wavelengths much longer than the computational domain, the quasi-static approximation is applicable. Under this condition, the electric potential V can be computed by solving the generalized Laplace equation:

$$\nabla \cdot (\sigma \nabla V) = 0 \quad \text{in } \Omega \tag{1}$$

where σ denotes the electrical conductivity, which depends on the type of tissue (tumor or normal) and the local temperature, and Ω represents the analysis domain. The conductivity increases linearly by 2% per degree Celsius as the temperature increases, but drops to zero when the temperature exceeds 100 °C due to tissue coagulation and carbonization [3,20]. The baseline conductivity values for normal and tumor tissues were taken from [26].

Equation 1 was numerically solved using the finite element method (FEM), with boundary conditions as defined:

$$\begin{cases} V(\mathbf{x}) = V_p, & \mathbf{x} \in \Gamma_e \\ \mathbf{n}(\mathbf{x}) \cdot \nabla V(\mathbf{x}) = 0, & \mathbf{x} \in \Gamma_s \end{cases} \tag{2}$$

where V_p denotes the applied voltage, Γ_e represents the electrode boundary, and Γ_s is the outer surface of the computational domain.

Bio-heat Propagation Analysis. The electric potential V is used to calculate the external heat source $Q_r = \sigma |\nabla V|^2$, which is incorporated into the Pennes bioheat model (Eq. 3) to model bioheat propagation within breast tissue [14].

$$\rho c \frac{\partial T}{\partial t} = \nabla \cdot (\kappa \nabla T) + \rho_b c_b \omega_b (T_b - T) + Q_m + Q_r \tag{3}$$

where T is the temperature, ρ the tissue density, c the specific heat, κ the thermal conductivity, and Q_m the metabolic heat. Blood-related terms include ρ_b, c_b, ω_b, and T_b, representing the density, specific heat, perfusion rate, and temperature of blood, respectively [13].

Equation 3 is solved using the finite difference time domain (FDTD) method [8] over 18 time steps (10 s each, total 180 s). After each step, the temperature-dependent conductivity is updated and Q_r is recalculated, enabling iterative coupling between electrical and thermal simulations.

Cell Necrosis Region Estimation. The temperature T obtained from the previous simulation is used in the Arrhenius model [1] (Eq. 4) to compute thermal damage $D(t)$:

$$D(t) = \int_0^t A \exp\left(\frac{-E_a}{RT(t)}\right) dt \tag{4}$$

where A is the frequency factor, E_a is the activation energy, R is the universal gas constant, and T is converted to absolute temperature (K). At each time step of the FDTD simulation, damage is accumulated by numerically integrating the model. Voxels with $D(t) > 1$ are classified as necrotic tissue, indicating the predicted ablation zone, with the threshold adopted from [6].

As a result, two types of simulation-based synthetic labels were generated across 18 time steps: tissue damage regions and temperature distributions. The damage regions are represented as 3D binary images indicating tumor occupancy, while the temperature distributions contain real temperature values (in °C) at each pixel. A total of 13 tumors were simulated, each with 500 different electrode placements, yielding 6,500 input-label pairs for each of the two prediction tasks.

2.3 Neural Network Model for Thermal Effect Prediction

Transformer-Based U-Net. A Transformer-based U-Net (UNETR) [5] was adopted to predict the thermal effects of RFA, as shown in the yellow box in Fig. 1. The choice of a Transformer-based encoder is closely tied to the characteristics of the input data. The electrode, one of the two input components, is a simple linear structure about 10mm in length and occupies a small region relative to the entire input domain. However, since thermal effects occur primarily around the electrode, its spatial location is crucial for accurate prediction. CNN-based models [12,15] may miss such spatial variations due to translation invariance, whereas Transformers preserve global positional information via attention and positional encoding [4]. Moreover, local features such as electrode orientation are effectively captured by the convolutional layers in the patch embedding module, contributing to more precise predictions of thermal effects.

The input is composed of two concatenated 3D volumes: a binary tumor geometry that provides material information by distinguishing tumor from normal tissue, and a needle geometry that provides the spatial location where ablation occurs. The output is a 18-channel 3D volume that represents predictions in 18 time steps. Two identical networks were trained separately for temperature and damage prediction, each with different target labels.

The input volumes first pass through a patch embedding module, where they are divided into non-overlapping patches. Each patch is embedded into a high-dimensional token vector with added positional encoding. These tokens are processed through 12 Transformer blocks (see Fig. 1 for detailed architecture), and hidden states are extracted from the 3rd, 6th, 9th, and 12th layers. The decoder follows a U-Net-like structure. The extracted hidden states are merged with decoder features via skip connections and progressively upsampled to recover the original spatial resolution of the input.

Multi-channel Outputs and Loss Functions. We predict all 18 time steps simultaneously for three key reasons: (1) Efficiency: Predicting each step separately requires 18 forward passes, increasing computation. A single 18-channel output enables fast, parallel inference. (2) Temporal continuity: Sharing a feature map across all time steps allows the model to implicitly learn temporal consistency, embedding smooth and coherent evolution patterns over time. (3) Temporal regularization: Loss is computed over the entire 18-channel volume rather than per time step, which encourages temporal regularization. This leads to more stable and reliable predictions. We use MSE loss for temperature prediction and Dice loss for damage area prediction.

3 Results

3.1 Dataset Composition

Since digital twin simulations require patient-specific RFA treatment planning, achieving high prediction performance on previously unseen tumor geometries is a critical requirement. To ensure a sufficient amount of training data given the limited variety of tumor shapes, a total of 5,500 samples (per task) were extracted from 11 out of 13 tumors and used for training. Among these, data from 9 tumors were used to train the model, and data from 2 tumors were used for validation with the early stopping technique. The remaining 1,000 samples from the 2 unseen tumors were used to evaluate the generalization performance of the model. The test set consisted of two subsets: one was a foreseen dataset, consisting of 1,000 samples randomly drawn from the training data; the other was an unforeseen dataset, constructed from the two unseen tumors.

3.2 Model Performance and Comparison

Predictive Accuracy. To evaluate the predictive accuracy of the proposed framework and to validate the appropriateness of the model selection, a comparative analysis was conducted with a CNN-based model. Table 1 presents the performance evaluation results for two tasks—temperature distribution prediction and damage area prediction—using both the foreseen and unforeseen datasets, in comparison with the CNN-based model. All evaluation metrics were computed over the entire set of 18 output channels, consistent with the training configuration, rather than being averaged over individual channels. In the temperature distribution prediction task, the transformer-based U-Net was the only model to achieve an RMSE below 1°C on both datasets, demonstrating its superior regression performance. In the damage area prediction task, it also achieved the highest accuracy across both Dice and IoU metrics, confirming the task suitability of the transformer-based model.

Temporal Prediction Stability. To enable stable simulations applicable to tumors of varying sizes, it is important that all time steps yield consistently accurate predictions. As shown in Fig. 3(a), the MSE loss trend across time steps reveals that both U-Net and Attention U-Net diverge toward unacceptable error levels at later steps. Since the affected region in the early steps is relatively small, the error tends to increase over time for all models. Although the

Table 1. Predictive performance on temperature distribution and damage area across models for foreseen and unforeseen dataset

Dataset	Model	Temperature Distribution				Damage Area	
		MSE	RMSE	MAE	PCC	Dice	IoU
Foreseen	UNETR	0.2853	0.5304	0.5765	0.9985	0.9766	0.9508
	Att. U-Net	4.4012	2.0632	0.9927	0.9833	0.9449	0.8971
	U-Net	1.2907	1.1085	0.7786	0.9920	0.9398	0.8870
Unforeseen	UNETR	0.5474	0.7160	0.6767	0.9975	0.9438	0.8948
	Att. U-Net	4.9328	2.1386	1.0321	0.9786	0.9137	0.8427
	U-Net	2.2620	1.4707	0.9061	0.9881	0.9027	0.8250

transformer-based U-Net also exhibits a rising loss trend, it maintains relatively stable predictive performance, with the final time step yielding a loss value close to 1. A visual comparison in Fig. 2(a) further confirms that, unlike the CNN-based models, which produce fragmented and noisy outputs, the transformer-based model effectively predicts smooth and coherent distributions that closely resemble the synthetic labels.

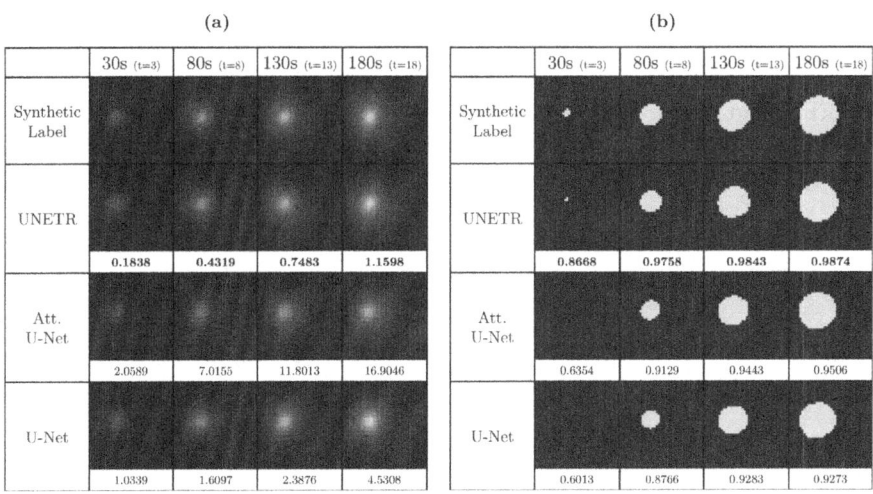

Fig. 2. Prediction results of each model on the unforeseen data at 30, 80, 130, and 180 s (a) Predicted temperature distributions and corresponding MSE loss (b) Predicted damage areas and corresponding Dice scores

Figure 2(b) shows the damage areas predicted by each model. In cases with small ablation regions, CNN-based models either fail to detect the region entirely or produce results smaller than the synthetic labels. In contrast, the transformer-based model accurately predicts not only the size but also the morphological details of the label, except in very small regions. As shown in Fig. 3(b), while the performance of CNN-based models significantly degrades at earlier time steps, the transformer-based model maintains relatively consistent temporal prediction stability, achieving Dice scores above 0.9 across all time steps.

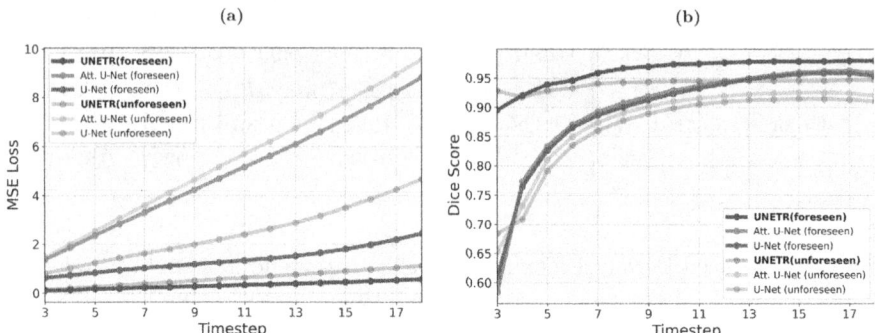

Fig. 3. Comparison of performance across time steps: (a) uses MSE, while (b) uses the Dice score

3.3 Inference Time

Table 2 compares the inference time of each method across all time steps in the RFA simulation. The transformer-based model, though slower than CNN-based models, offers higher accuracy and drastically reduces computation time compared to the numerical method. CNN-based models are faster due to smaller sizes but have lower accuracy. All inference time measurements were conducted on a desktop computer utilizing GPU computation. (AMD Ryzen 9 5900X 12-Core Processor 3701Mhz and NVIDIA GeForce RTX 3090).

Table 2. Comparison of numerical simulation and deep learning inference time for 18-channel output

	Numerical Sim.	UNETR	Att. U-Net	U-Net
Time (s)	**76.2268**	**0.0468**	0.0048	0.0025

4 Conclusion

This study proposes a neural network model capable of real-time simulation of the thermal effects of RF ablation in a virtual environment. By reducing the long inference time caused by the high computational cost of traditional numerical simulations by a factor of approximately 1,600, the model accurately predicts temperature distribution and tissue damage throughout the entire procedure, demonstrating its potential as an alternative to conventional simulation methods. As interest in digital twin-based personalized treatment and diagnosis continues to grow, the proposed model has shown fast and accurate performance across various tumor geometries. Future research will incorporate patient-specific organ characteristics—such as blood vessels, fat, and fibrosis—instead of relying on generalized biophysical parameters, and will expand the tumor dataset to address morphological limitations, aiming to enable clinically applicable treatment planning.

Acknowledgments. The work was supported by the National Research Foundation of Korea (NRF) funded by the Korean government (MSIT) under Grants RS-2024-00335185 and RS-2023-00220762.

Disclosure of Interests. All authors declare that they have no known competing financial interests or personal relationships that may influence the work reported in this paper.

References

1. Arrhenius, S.: Über die reaktionsgeschwindigkeit bei der inversion von rohrzucker durch säuren. Z. Phys. Chem. **4**(1), 226–248 (1889)
2. Burger, W., Burge, M.J., Burge, M.J., Burge, M.J.: Principles of Digital Image Processing, vol. 111. Springer (2009)
3. Dodde, R.E., Miller, S.F., Geiger, J.D., Shih, A.J.: Thermal-electric finite element analysis and experimental validation of bipolar electrosurgical cautery (2008). https://doi.org/10.1115/1.2902858
4. Dosovitskiy, A., et al.: An image is worth 16x16 words: transformers for image recognition at scale. arXiv preprint arXiv:2010.11929 (2020). https://doi.org/10.48550/arXiv.2010.11929
5. Hatamizadeh, A., et al.: Unetr: transformers for 3d medical image segmentation. In: Proceedings of the IEEE/CVF Winter Conference on Applications of Computer Vision, pp. 574–584 (2022)
6. Henriques, F.C., Jr.: Studies of thermal injury; the predictability and the significance of thermally induced rate processes leading to irreversible epidermal injury. Arch. Pathol. **43**(5), 489–502 (1947)
7. Jiang, Y., et al.: Formulation of 3D finite elements for hepatic radiofrequency ablation. Int. J. Model. Ident. Control **9**(3), 225–235 (2010). https://doi.org/10.1504/IJMIC.2010.032803
8. Kim, H.J., et al.: Laser-tissue interaction simulation considering skin-specific data to predict photothermal damage lesions during laser irradiation. J. Comput. Des. Eng. **10**(3), 947–958 (2023). https://doi.org/10.1093/jcde/qwad033
9. Lee, D.H., Lee, J.M.: Recent advances in the image-guided tumor ablation of liver malignancies: radiofrequency ablation with multiple electrodes, real-time multi-modality fusion imaging, and new energy sources. Korean J. Radiol. **19**(4), 545–559 (2018). https://doi.org/10.3348/kjr.2018.19.4.545
10. Mariappan, P., et al.: GPU-based RFA simulation for minimally invasive cancer treatment of liver Tumours. Int. J. Comput. Assist. Radiol. Surg. **12**, 59–68 (2017). https://doi.org/10.1007/s11548-016-1469-1

11. McDermott, S., Gervais, D.A.: Radiofrequency ablation of liver tumors. Seminars Intervent. Radiol. **30**, 049–055. Thieme Medical Publishers (2013). https://doi.org/10.1055/s-0033-1333653
12. Oktay, O., et al.: Attention u-net: learning where to look for the pancreas. arXiv preprint arXiv:1804.03999 (2018). https://doi.org/10.48550/arXiv.1804.03999
13. Paruch, M.: Mathematical modeling of breast tumor destruction using fast heating during radiofrequency ablation. Materials **13**(1), 136 (2019). https://doi.org/10.3390/ma13010136
14. Pennes, H.H.: Analysis of tissue and arterial blood temperatures in the resting human forearm. J. Appl. Physiol. **1**(2), 93–122 (1948). https://doi.org/10.1152/jappl.1948.1.2.93
15. Ronneberger, O., Fischer, P., Brox, T.: U-Net: convolutional networks for biomedical image segmentation. In: Navab, N., Hornegger, J., Wells, W.M., Frangi, A.F. (eds.) MICCAI 2015. LNCS, vol. 9351, pp. 234–241. Springer, Cham (2015). https://doi.org/10.1007/978-3-319-24574-4_28
16. Saha, A., et al.: A machine learning approach to radiogenomics of breast cancer: a study of 922 subjects and 529 DCE-MRI features. Br. J. Cancer **119**(4), 508–516 (2018). https://doi.org/10.1038/s41416-018-0185-8
17. Servin, F., et al.: Simulation of image-guided microwave ablation therapy using a digital twin computational model. IEEE Open J. Eng. Med. Biol. **5**, 107–124 (2023). https://doi.org/10.1109/OJEMB.2023.3345733
18. Shin, M., et al.: Physrfanet: physics-guided neural network for real-time prediction of thermal effect during radiofrequency ablation treatment. Eng. Appl. Artif. Intell. **138**, 109349 (2024). https://doi.org/10.1016/j.engappai.2024.109349
19. Shin, M., Seo, M., Yoo, S.S., Yoon, K.: tFUSFormer: physics-guided super-resolution transformer for simulation of transcranial focused ultrasound propagation in brain stimulation. IEEE J. Biomed. Health Inform. (2024). https://doi.org/10.1109/JBHI.2024.3389708
20. Singh, S., Repaka, R.: Parametric sensitivity analysis of critical factors affecting the thermal damage during RFA of breast tumor. Int. J. Therm. Sci. **124**, 366–374 (2018). https://doi.org/10.1016/j.ijthermalsci.2017.10.032
21. Sun, T., He, X., Li, Z.: Digital twin in healthcare: recent updates and challenges. Digital Health **9**, 20552076221149652 (2023). https://doi.org/10.1177/20552076221149651
22. Tatli, S., Tapan, Ü., Morrison, P.R., Silverman, S.G.: Radiofrequency ablation: technique and clinical applications. Diagn. Interv. Radiol. **18**(5), 508 (2012). https://doi.org/10.4261/1305-3825.DIR.5168-11.1
23. Tungjitkusolmun, S., et al.: Three-dimensional finite-element analyses for radio-frequency hepatic tumor ablation. IEEE Trans. Biomed. Eng. **49**(1), 3–9 (2002)
24. Vallée, A.: Digital twin for healthcare systems. Front. Digital Health **5**, 1253050 (2023). https://doi.org/10.3389/fdgth.2023.1253050
25. Voglreiter, P., et al.: RFA guardian: comprehensive simulation of radiofrequency ablation treatment of liver tumors. Sci. Rep. **8**(1), 787 (2018). https://doi.org/10.1038/s41598-017-18899-2
26. Zhao, Z., et al.: System development of microwave induced thermo-acoustic tomography and experiments on breast tumor. Progress Electromagn. Res. **134**, 323–336 (2013)
27. Zorbas, G., Samaras, T.: Simulation of radiofrequency ablation in real human anatomy. Int. J. Hyperth. **30**(8), 570–578 (2014). https://doi.org/10.3109/02656736.2014.968639

Finite-Element Electrophysiological Modeling of Human Uterine Smooth Muscle Using a Reduced Tong Model

Zhen Li and Alberto Corrias(✉)

Department of Biomedical Engineering, National University of Singapore,
Blk E4, #04-08, 4 Engineering Drive 3, Singapore 117583, Singapore
bieac@nus.edu.sg

Abstract. The uterus is a central organ in the female reproductive system, yet its electrophysiological mechanisms remain poorly understood due to its dynamic functional states and the scarcity of human data. Computational modeling has thus emerged as one of the most effective approaches to studying excitation–contraction mechanisms under constrained experimental conditions. This study presents a finite-element electrophysiological model of uterine smooth muscle cells (uSMCs) based on the Reduced Tong Model (RTM). The proposed framework more accurately captures key physiological behaviors of the uterus while reducing computational cost. Using this model, we experimentally determined that a tissue conductivity between 0.02 and 0.30 mS/cm supports physiologically realistic conduction velocities in uterine tissue. We further performed a qualitative investigation of oxytocin-induced excitability changes and proposed a potential dual role of intracellular calcium in both facilitating and suppressing the initiation of action potentials.

Keywords: Uterine Electrophysiology · Finite Element Modeling · Oxytocin

1 Introduction

The uterus is a central organ in the female reproductive system, yet its electrophysiological mechanisms remain far less studied compared to those of the heart and gastrointestinal system [1, 2]. This disparity largely stems from the uterus's uniquely dynamic physiological behavior. The uterine contraction pattern differs markedly between pregnant and non-pregnant states and continues to evolve throughout pregnancy [3, 4]. Such frequent shifts in functional state are rare among human organs and pose considerable challenges to understanding the mechanisms that initiate and maintain excitation–contraction coupling [5].

Despite decades of research, how the uterus generates pacemaker activity during labor and coordinates coherent contractions remains incompletely understood. Unlike the heart and gastrointestinal tract - whose rhythmic activities are governed by specialized centers such as the sinoatrial node [2] and interstitial cells of Cajal (ICCs) [6] - the uterus appears to lack a clearly defined pacemaker region [7]. Instead, studies suggest that uterine action potentials may originate from multiple, spatially dispersed loci [8]. This combination of spatial heterogeneity and temporally dynamic functional states makes the uterus one of the most complex rhythmically excitable organs in the human body [9].

One of the main barriers to studying uterine electrophysiology is the severe lack of human data, as ethical constraints make high-resolution, invasive measurements virtually impossible [10, 11]. As a result, most current knowledge is derived from animal studies. However, uterine electrical behavior is highly state-dependent, and marked discrepancies exist between in vivo and in vitro conditions [12, 13]. Furthermore, interspecies differences in gestational physiology, myometrial structure, and ion channel expression [14, 15] further limit the direct applicability of animal-derived data to human uterine research [16, 17].

To address these challenges, the development of multi-scale, high-fidelity, and interpretable computational models has emerged as one of the most promising approaches to advancing our understanding of uterine electrophysiology [18–21]. However, many existing studies remain limited to the single-cell level, lacking the research to simulate tissue- and organ-level behavior [22], or are based on overly simplified cellular models that fail to capture the full complexity of uterine smooth muscle cell (uSMC) physiology [23, 24].

To overcome these limitations, we developed a multi-scale finite element model of uterine electrophysiology. This model simulates the electrical activity of uterine smooth muscle cells (uSMCs) and its spatial propagation across the tissue. The model is based on the single-cell electrophysiological framework proposed by Tong et al. [25] and is implemented in Chaste [26], a widely used FEM simulation platform for cardiac electrophysiology. Using this model, we explored a physiologically plausible range of tissue conductance values and qualitatively reproduced oxytocin-induced excitability enhancement, demonstrating the tissue's increased likelihood of initiating contractions in response to stimulation.

2 Method

As noted above, previous uterine modeling studies have achieved significant progress but still have notable limitations. The multiscale framework developed by Yochum et al. [20, 23] remains one of the most comprehensive and influential models to date, successfully extending simulations to three dimensions and reproducing many physiological phenomena. However, this model has two key drawbacks: first, it relies on the overly simplified Rihana single-cell model [18] due to computational constraints at the time, which limits its ability to capture detailed calcium dynamics and force generation; second, it uses a traditional cell-to-cell coupling scheme instead of a finite element method, resulting in higher computational cost and reduced flexibility for representing tissue heterogeneity, anisotropy, and realistic boundaries.

Xu's study [32] adopted the more detailed Tong single-cell model [25] and implemented a simple heterogeneous tissue grid in two dimensions, but also did not use finite element methods, leading to limitations in handling complex spatial structures, anisotropy, and organ-scale simulations.

Building on these previous studies, we adopted a multiscale modeling strategy that balances biophysical realism and computational efficiency, with two main improvements: updating the single-cell model and employing the FEM at the tissue level.

The single-cell model used here is the Reduced Tong Model (RTM) by Means et al. [22]. The RTM retains key electrophysiological mechanisms, such as L-type and T-type calcium channels and BK potassium channels. Non-essential currents, including K2 and Ka channels, were removed to simplify the RTM. The number of differential equations was reduced from 19 to 5.

$$\frac{dV_m}{d_t} = -(I_{Na} + I_{CaL} + I_{CaT} + I_{Cl} + I_{K1} + I_{BK} + I_{NS} + I_b + I_{stim}) \tag{1}$$

The ionic current through channel x is described using a standard Hodgkin–Huxley-type formulation, and the gating variables evolve according to first-order kinetics:

$$I_x = g_{\infty,x} m_x h_x (V_m - E_x) \tag{2}$$

$$\frac{dm}{d_t} = \frac{m_\infty(V_m) - m}{\tau_m(V_m)} \tag{3}$$

In this expression, m denotes the instantaneous value of the activation variable, $m_\infty(V_m)$ is its steady-state value at a given membrane voltage, and $\tau_m(V_m)$ is the voltage-dependent relaxation time constant (in ms). In the RTM, gating variables with fast kinetics $(\tau_m(V_m) \ll \Delta t)$[1] are set to their steady-state values, eliminating the need to solve their differential equations and greatly reducing computational cost while preserving essential electrophysiological behavior.

The RTM achieves a 2.6-fold speed-up over the full Tong model (FTM) in 2D simulations and a 3.7-fold speed-up in 3D [22]. Given these advantages, the RTM model offers a high-fidelity and computationally efficient single-cell foundation for tissue-level finite element modeling of uterine electrical activity.

The RTM was adapted for use within the Chaste framework through some structural adjustments and the addition of RDF metadata annotations. Then a monodomain finite element formulation was adopted to simulate tissue-level electrophysiological dynamics, as described by Eq. (4) [28]:

$$\frac{dV}{dt} = \nabla \cdot (D\nabla V) - \frac{I_{ion}}{C_m} \tag{4}$$

As illustrated in Fig. 1, tissue-scale simulations were performed using a two-dimensional finite element grid consisting of 51 × 51 nodes with a uniform spacing of 0.01 cm, representing a square tissue domain. Electrical stimulation was applied to a 9 × 9 region in the lower-left corner. Each node in the grid is assigned an RTM, and electrical propagation throughout the tissue is governed by the finite element discretization of the monodomain equation, which captures both local cellular dynamics and spatial voltage diffusion.

---

[1] Δt Is the simulation time step used in numerical integration (typically around 10 ms).

The key distinction from the traditional cell-to-cell model is that the finite element approach treats electrical propagation as a continuous process governed by tissue conductivity, rather than discrete resistive coupling between individual cells. This continuous formulation provides greater flexibility for extending the model to more complex tissue structures and larger spatial scales.

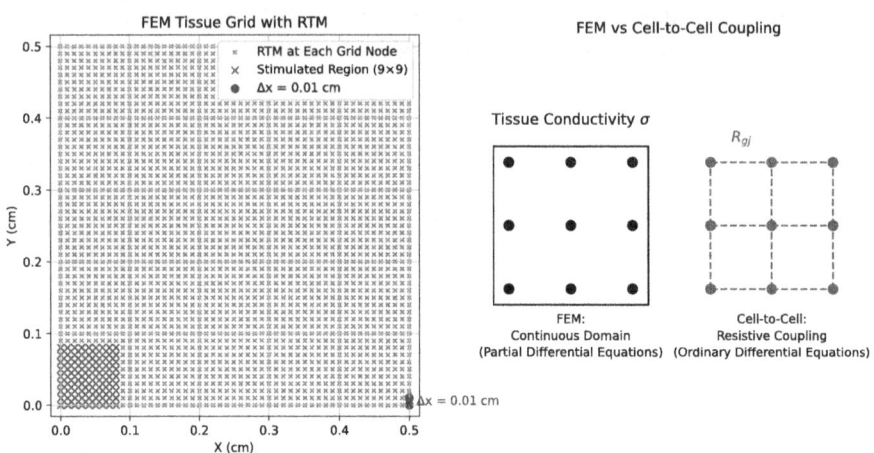

Fig. 1. Comparison of FEM and cell-to-cell modeling for tissue electrophysiology. Left: FEM grid with RTM in our model. Right: A sample of FEM uses tissue conductivity σ, while cell-to-cell coupling uses gap junction resistance R_{gj}.

All simulations were performed on an Ubuntu (VMware) virtual machine hosted on Windows, equipped with a 4-core Intel 11th Gen Core i7-11800H CPU. On a 51 × 51 finite element grid, each 1000 ms simulation required approximately 20 min of computation time.

3 Experiment

3.1 Construction of a Physiologically Realistic FEM Model

Our primary goal is to construct a multiscale finite element model that couples a biophysically accurate single-cell model with a flexible spatial discretization scheme. This framework, extensively validated in cardiac electrophysiology, enables high-fidelity simulation of tissue-level electrical activity.

We simulated a 1 cm^2 anisotropic uterine tissue, with transverse conductivity set to 0.120 mS/cm and longitudinal conductivity to 0.020 mS/cm (as detailed in Sect. 3.2), thereby qualitatively capturing the faster propagation of electrical signals along the muscle fiber direction.

Figure 2 illustrates the detailed electromechanical behavior observed in the model. Upon stimulation at the lower-left corner, an elliptically shaped wavefront emerges, reflecting anisotropic excitation governed by the tissue conductivities. Intracellular calcium concentration rises rapidly following membrane depolarization, producing a sharp and immediate peak. However, its return to baseline is substantially slower than the repolarization of the membrane potential. This physiologically realistic temporal dissociation differs from the results reported in Yochum's model [20, Fig. 3], which lacked explicit calcium dynamics.

In contrast to the rapid changes in voltage and calcium, active force generation exhibits a marked delay. Due to the slow decay of intracellular calcium, force accumulates gradually through the cross-bridge mechanism and persists well beyond the electrical event. In corresponding single-cell simulations, we observed that force peaks approximately 2000 ms after a single action potential and then decays slowly.

This delayed mechanical response highlights a key physiological distinction of uterine smooth muscle. Unlike other muscle types, contraction does not terminate immediately after repolarization. Instead, the muscle continues to generate force as long as intracellular calcium concentration remains at a relatively high level, even though it is gradually declining. This property enables the uterus to maintain tone or produce sustained contractions required during labor.

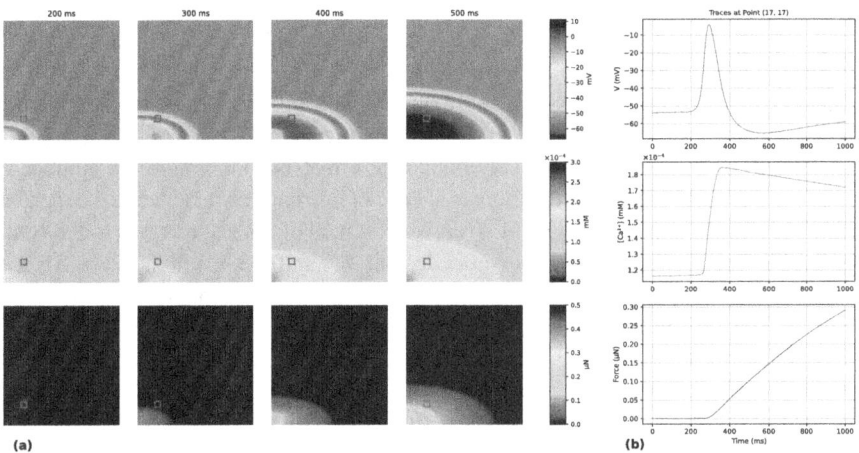

Fig. 2. Spatiotemporal evolution of transmembrane voltage, intracellular calcium concentration, and active force in a 2D homogeneous uterine tissue model. (a) Snapshots at successive time points show that electrical excitation propagates rapidly following stimulation. Calcium concentration increases shortly after depolarization, but returns to baseline much more slowly. Active force development further lags behind calcium. (b) Time traces at a representative point highlight the temporal dissociation between electrical activity, calcium dynamics, and mechanical response.

3.2 Calibration of Tissue Conductivity

As previously noted, physiological measurements of uterine tissue conductivity are extremely limited due to ethical constraints. This scarcity of empirical data makes it difficult to determine appropriate conductance values for simulation, despite their critical role in electrical propagation. Moreover, the conductance values used in existing studies vary considerably [22, 24], further highlighting the need for systematic calibration.

To address this uncertainty, we systematically varied the tissue conductivity parameter in our 2D model and measured the corresponding conduction velocities. By comparing these results with experimentally reported physiological conduction velocities, we identified a plausible range of conductance values suitable for tissue-level simulations.

From a theoretical perspective, both excessively low and excessively high tissue conductance can lead to propagation failure, albeit through different mechanisms. Low conductance weakens electrotonic coupling, making it difficult for adjacent cells to reach the depolarization threshold. In contrast, overly high conductance causes the current to spread too rapidly, reducing local current density and thus impairing downstream excitation. Therefore, the conductance values used in the uterine model must fall within a range that is both numerically feasible and physiologically plausible, ensuring stable propagation while remaining consistent with experimentally observed conduction velocities.

To ensure accurate measurement of conduction velocity, we selected two points along the diagonal of an isotropic tissue domain and calculated the time delay between their respective peak depolarizations. However, as shown in Fig. 3a, the local propagation speed increases sharply near the edges and corners of the tissue. This effect arises from the sealed, non-conductive boundary conditions used in the simulation, which are unavoidable in finite-domain modeling.

When the wavefront approaches these boundaries, electrical current cannot exit the domain, leading to an artificial rise in local membrane potential and a non-physiological increase in conduction velocity. This boundary effect becomes more pronounced at higher conductance values, as the increased ease of current flow exacerbates edge accumulation. To eliminate this source of error, we confined our measurements to the central region of the grid, specifically between nodes (17,17) and (33,33) in the 51 × 51 grid.

The measured conduction velocities under varying tissue conductance are presented in Fig. 3b. Two sets of simulations were conducted using different L-type calcium channel conductance values ($gCaL = 0.6$ and $0.7 nS/pF$). In both cases, the relationship between longitudinal tissue conductance and conduction velocity followed a nonlinear trend that was well approximated by a square-root function of the form $v = a\sqrt{g} + b$. This relationship is theoretically supported by cable theory, which predicts that conduction velocity scales with the square root of axial conductance under the assumption of constant membrane properties [30].

Based on limited physiological experimental evidence, the expected conduction velocity in human uterine smooth muscle ranges from approximately 0.001 to 0.004 cm/ms [13, 29]. When using the original RTM single-cell formulation (with $gCaL = 0.6nS/pF$), this physiological range corresponds to a tissue conductance between roughly 0.02 and $0.30 mS/cm$. It is worth noting that conduction failure occurred even before the simulated velocity reached the upper bound of the physiological range.

To qualitatively simulate the excitability-enhancing effects of hormones such as estrogen and oxytocin, the L-type calcium channel conductance was increased to $0.7 nS/pF$. Under this condition, the tissue exhibited conduction velocities exceeding $0.003 cm/ms$, consistent with values reported in physiological experiments. These results align with the known role of hormonal influences during late pregnancy and labor, where agents are believed to increase uterine excitability and facilitate faster propagation of electrical signals.

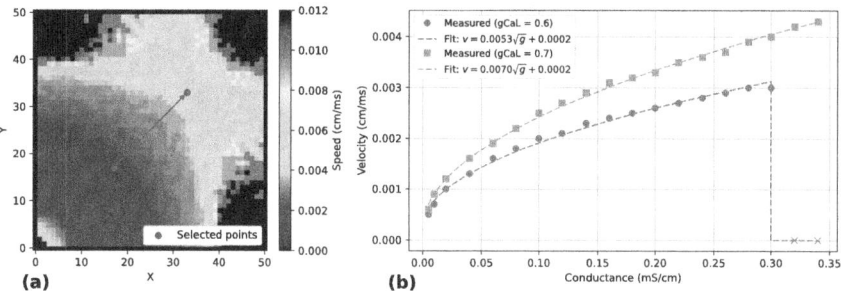

Fig. 3. Relationship between tissue conductance and conduction velocity. (a) Local conduction velocity map in 2D uterine tissue. (b) Velocity vs. longitudinal conductance for two gCaL values (0.6, 0.7 nS/pF). Experimental data (dots) are fitted using a square-root function. Red crosses indicate conduction failure.

3.3 Simulation of Oxytocin-Induced Excitability Changes

Based on the preceding results, we conducted a preliminary simulation to explore the excitability-enhancing effects of oxytocin. As shown in Fig. 4, a weak periodic stimulus of $-0.12 pA/pF$ was applied to the tissue, and the resulting electrophysiological and mechanical responses were compared between the non-hormonal condition (Fig. 4a–e) and the simulated oxytocin condition (Fig. 4f–j).

Physiologically, oxytocin primarily increases cellular excitability by regulating IP_3R [31]. However, because IP_3R is not currently incorporated in any available uterine cell models, we instead approximated the hormonal effect by increasing the L-type calcium channel conductance gCaL as a qualitative surrogate.

Under non-hormonal conditions, the initial stimuli failed to trigger action potentials (Fig. 4b). Nevertheless, with progressive accumulation of intracellular calcium, successful excitation occurred in the third cycle. This may suggest that elevated $[Ca^{2+}]$ enhances the activation of L-type Ca^{2+} channels, facilitating subsequent depolarization and action potential generation.

At the same time, elevated calcium levels may trigger inhibitory feedback mechanisms. In the baseline case, brief conduction failure is observed during the fourth cycle, and similar suppression occurs intermittently after each action potential under oxytocin conditions. We speculate that this effect is mediated by the BKab current—a calcium-activated potassium current known to suppress excitability. As shown in Fig. 4e and 3j, noticeable BKab fluctuations occur even under subthreshold stimulation, indicating that excessive intracellular $[Ca^{2+}]$ can inhibit action potential firing by activating hyperpolarizing outward K^+ currents.

This inhibitory effect also prevents unbounded force accumulation. As shown in Fig. 4d and 3i, despite continuous stimulation, the uterine smooth muscle produces a relatively stable and plateau-like active force. Such regulation is physiologically desirable, allowing the uterus to maintain effective tone or sustained contractions during labor without entering a state of pathological over-contraction.

Taken together, these observations highlight the dual role of intracellular calcium: moderate elevation promotes excitability, while both insufficient and excessive levels may impair action potential initiation.

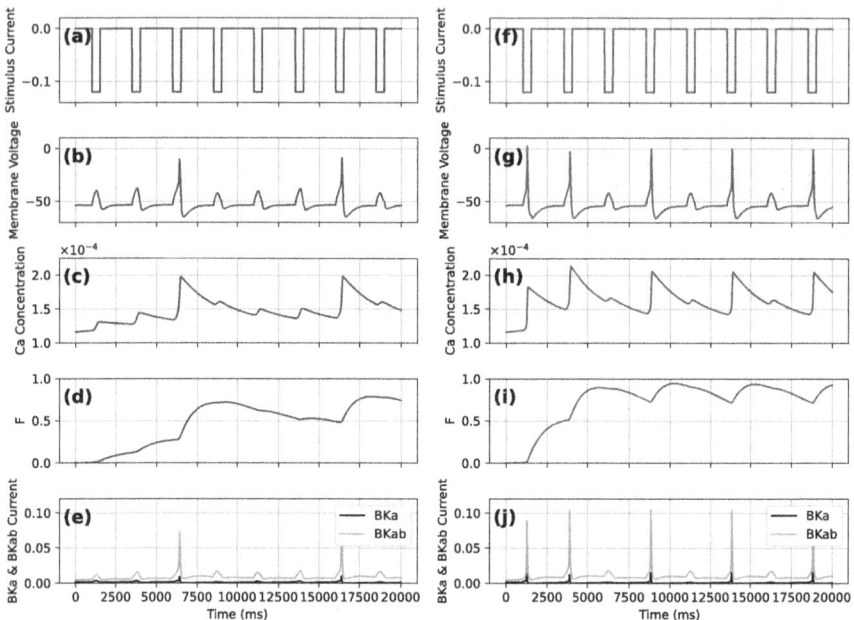

Fig. 4. Simulation of excitability modulation via L-type calcium conductance. Left: baseline ($gCaL = 0.6nS/pF$). Right: oxytocin-simulated ($gCaL = 0.7nS/pF$). Panels show stimulus (a,f), voltage (b,g), calcium (c,h), force (d,i), and BKa/BKab currents (e,j) over 20 s.

4 Conclusions

This study developed and validated a finite-element electrophysiological model of uterine smooth muscle cells (uSMCs) based on the Reduced Tong Model, providing a platform for studying excitation under specific simulated conditions. The proposed model overcomes key limitations of previous frameworks by integrating a biophysically detailed yet computationally efficient single-cell model with a flexible finite element method for tissue-level simulation. This approach more accurately reproduces key physiological behaviors, including the spatiotemporal dynamics of voltage, calcium, and force.

Using this model, we identified a plausible range of tissue conductivities (0.02–0.30 mS/cm) supporting physiologically realistic conduction velocities, providing a valuable reference point for future uterine electrophysiology simulations. We also qualitatively simulated oxytocin-induced excitability changes by modulating L-type calcium channel conductance and proposed a dual role for intracellular calcium: moderate elevation promotes excitability, while both insufficient and excessive levels may impair action potential initiation.

It is worth emphasizing that the current research remains at a nascent stage across multiple levels. First, the model is currently confined to the two-dimensional tissue level. However, leveraging Chaste's robust architecture, extension to three dimensions is readily achievable given well-defined mesh files; transition between monodomain and bidomain formulations is also possible with minor code modifications.

Second, this study has preliminarily demonstrated the model's capability to incorporate anisotropy and heterogeneity, future work will focus on systematically defining these parameters within a three-dimensional organ-scale framework. This will enable simulations investigating physiological differences in uterine electrophysiology and contractility across distinct pregnancy stages and phases of labor under hormonal influences.

Third, while RTM offering computational efficiency, still presents significant opportunities for enhancement. Notably, it currently lacks mechanisms for intrauterine pressure feedback to modulate electrical activity and omits certain ion channels. These omissions limit the ability to quantitatively define hormonal effects with higher precision. Incorporating these essential physiological mechanisms, while maintaining computational tractability, is highly valuable for achieving a more accurate representation of uterine phenomena.

To sum up, this work establishes a foundational computational framework for uterine electrophysiology. Addressing these limitations represents critical next steps toward developing a comprehensive, multi-scale model capable of simulating the dynamic physiological states characteristic of human uterine function.

Acknowledgments. We thank Assistant Professor Lei Li (National University of Singapore), PhD student Yilin Lyu, and PhD student Fan Yang (Sichuan University) for their generous assistance during the writing of this paper.

Disclosure of Interests. The authors have no competing interests to declare that are relevant to the content of this article.

References

1. Rabotti, C., Mischi, M.: Propagation of electrical activity in uterine muscle during pregnancy: a review. Acta Physiol. **213**(2), 406–416 (2015)
2. Smith, R., Imtiaz, M., Banney, D., Paul, J.W., Young, R.C.: Why the heart is like an orchestra and the uterus is like a soccer crowd. Am. J. Obstet. Gynecol. **213**(2), 181–185 (2015)
3. van Gestel, I., IJland, M.M., Hoogland, H.J., Evers, J.L.H.: Endometrial wave-like activity in the non-pregnant uterus. Hum. Reprod. Update **9**(2), 131–138 (2003)
4. Bakker, P.C.A.M., Van Rijsiwijk, S., van Geijn, H.P.: Uterine activity monitoring during labor. J. Perinat. Med. **35**(6), 468–477 (2007)
5. Wray, S., Prendergast, C.: The myometrium: from excitation to contractions and labour. In: Smooth Muscle Spontaneous Activity, vol. 1124, H. Hashitani and R. J. Lang, Eds., Singapore: Springer Singapore, pp. 233–263 (2019)
6. Sanders, K.M.: A case for interstitial cells of Cajal as pacemakers and mediators of neurotransmission in the gastrointestinal tract. Gastroenterology (New York, N.Y. 1943) **111**(2), 492–515 (1996)
7. Lammers, W.J.E.P.: The electrical activities of the uterus during pregnancy. Reprod. Sci. **20**(2), 182–189 (2013)

8. Lammers, W.J.E.P., Stephen, B., Al-Sultan, M.A., Subramanya, S.B., Blanks, A.M.: The location of pacemakers in the uteri of pregnant guinea pigs and rats. Am. J. Physiol. Regul. Integr. Comp. Physiol. **309**(11), R1439–R1446 (2015)
9. Lammers, W.J.E.P., et al.: Patterns of electrical propagation in the intact pregnant guinea pig uterus. Am. J. Physiol. Regul. Integr. Comp. Physiol. **294**(3), R919–R928 (2008)
10. Rabotti, C., Mischi, M., van Laar, J.O.E.H., Oei, G.S., Bergmans, J.W.M.: Estimation of internal uterine pressure by joint amplitude and frequency analysis of electrohysterographic signals. Physiol. Meas. **29**(7), 829–841 (2008)
11. Xu, J., Chen, Z., Lou, H., Shen, G., Pumir, A.: Review on EHG signal analysis and its application in preterm diagnosis. Biomed. Signal Process. Control **71**, 103231 (2022)
12. Kao, C.Y.: Long-term observations of spontaneous electrical activity of the uterine smooth muscle. Am. J. Physiol. **196**(2), 343–350 (1959)
13. Verhoeff, A., Garfield, R.E., Ramondt, J., Wallenburg, H.C.S.: Electrical and mechanical uterine activity and gap junctions in peripartal sheep. Am. J. Obstet. Gynecol. **153**(4), 447–454 (1985)
14. Mitchell, B.F., Taggart, M.J.: Are animal models relevant to key aspects of human parturition? Am. J. Phys. Regul. Integr. Comp. Physiol. **297**(3), R525-545 (2009)
15. Malik, M., Roh, M., England, S.K.: Uterine contractions in rodent models and humans. Acta Physiologica **231**(4), e13607-n/a (2021)
16. Jain, V., Saade, G.R., Garfield, R.E.: Structure and function of the myometrium. In: Advances in Organ Biology, vol. 8, Elsevier B.V, pp. 215–246 (2000)
17. Weiss, S., et al.: Three-dimensional fiber architecture of the nonpregnant human uterus determined ex vivo using magnetic resonance diffusion tensor imaging. Anat. Record. Part A, Discoveries Mol. Cell. Evol. Biol. **288A**(1), 84–90 (2006)
18. Rihana, S., Terrien, J., Germain, G., Marque, C.: Mathematical modeling of electrical activity of uterine muscle cells. Med. Biol. Eng. Compu. **47**(6), 665–675 (2009)
19. Benson, A.P., Clayton, R.H., Holden, A.V., Kharche, S., Tong, W.C.: Endogenous driving and synchronization in cardiac and uterine virtual tissues: bifurcations and local coupling. Philos. Trans. R. Soc. London Ser. A: Math. Phys. Eng. Sci. **364**(1842), 1313–1327 (2006)
20. Yochum, M., Laforêt, J., Marque, C.: Multi-scale and multi-physics model of the uterine smooth muscle with mechanotransduction. Comput. Biol. Med. **93**, 17–30 (2018)
21. Sharifimajd, B., Thore, C.-J., Stålhand, J.: Simulating uterine contraction by using an electro-chemo-mechanical model. Biomech. Model. Mechanobiol. **15**(3), 497–510 (2016)
22. Means, S.A., Clark, A.R., Cheng, L.K.: Steady-state Approximations for Hodgkin-Huxley Cell Models: Towards Multi-scale Models of Uterine Smooth Muscle. Physiome (2023)
23. Yochum, M., Laforêt, J., Marque, C.: An electro-mechanical multiscale model of uterine pregnancy contraction. Comput. Biol. Med. **77**, 182–194 (2016)
24. Zahran, S.: Finite Element Modeling of Electrical Activity in Human Uterine Tissue: Advances in Simulation Techniques. Cold Spring Harbor Laboratory (2024)
25. Tong, W.-C., Choi, C.Y., Karche, S., Holden, A.V., Zhang, H., Taggart, M.J.: A computational model of the ionic currents, Ca.sup.2+ dynamics and action potentials underlying contraction of isolated uterine smooth muscle. PLoS One **6**(4), e18685 (2011)
26. Mirams, G.R., et al.: Chaste: an open source C++ library for computational physiology and biology. PLoS Comput. Biol. **9**(3), e1002970 (2013)
27. Tong, W.-C., Tribe, R.M., Smith, R., Taggart, M.J.: Computational modeling reveals key contributions of KCNQ and hERG currents to the malleability of uterine action potentials underpinning labor. PLoS ONE **9**(12), e114034 (2014)
28. Li, Q., Zhu, X., Chen, W.: Parallelization of three dimensional cardiac simulation on GPU. Biomedicines **12**(9), 2126 (2024)
29. Melton Carlton, E.J.R., Saldivar Julian, T.J.R.: Impulse velocity and conduction pathways in rat myometrium. Am. J. Physiol. **207**(2), 279–285 (1964)

30. Rall, W.: Theory of physiological properties of dendrites. Ann. N. Y. Acad. Sci. **96**(4), 1071–1092 (1962)
31. Wray, S., Burdyga, T.: Sarcoplasmic reticulum function in smooth muscle. Physiol. Rev. **90**(1), 113–178 (2010)
32. Xu, J., Menon, S.N., Singh, R., Garnier, N.B., Sinha, S., Pumir, A.: The role of cellular coupling in the spontaneous generation of electrical activity in uterine tissue. PLoS ONE **10**(3), e0118443–e0118443 (2015)

TF-TransUNet1D: Time-Frequency Guided Transformer U-Net for Robust ECG Denoising in Digital Twin

Shijie Wang and Lei Li(✉)

Department of Biomedical Engineering, National University of Singapore, Singapore, Singapore
lei.li@nus.edu.sgu

Abstract. Electrocardiogram (ECG) signals serve as a foundational data source for cardiac digital twins, yet their diagnostic utility is frequently compromised by noise and artifacts. To address this issue, we propose TF-TransUNet1D, a novel one-dimensional deep neural network that integrates a U-Net-based encoder–decoder architecture with a Transformer encoder, guided by a hybrid time–frequency domain loss. The model is designed to simultaneously capture local morphological features and long-range temporal dependencies, which are critical for preserving the diagnostic integrity of ECG signals. To enhance denoising robustness, we introduce a dual-domain loss function that jointly optimizes waveform reconstruction in the time domain and spectral fidelity in the frequency domain. In particular, the frequency-domain component effectively suppresses high-frequency noise while maintaining the spectral structure of the signal, enabling recovery of subtle but clinically significant waveform components. We evaluate TF-TransUNet1D using synthetically corrupted signals from the MIT-BIH Arrhythmia Database and the Noise Stress Test Database (NSTDB). Comparative experiments against state-of-the-art baselines demonstrate consistent superiority of our model in terms of SNR improvement and error metrics, achieving a mean absolute error of 0.1285 and Pearson correlation coefficient of 0.9540. By delivering high-precision denoising, this work bridges a critical gap in preprocessing pipelines for cardiac digital twins, enabling more reliable real-time monitoring and personalized modeling.

Keywords: ECG Denoising · Time Frequency · Deep Learning · Dual-Domain Loss · Cardiac Digital Twin

1 Introduction

Accurate electrocardiogram (ECG) signal denoising is a foundational prerequisite for cardiac digital twin systems, where high-fidelity physiological data drives personalized modeling and real-time diagnostics [1, 2]. While conventional signal processing techniques, such as linear filtering, Gaussian smoothing, and wavelet transforms [3], have been widely adopted for signal enhancement [4, 5], their rigidity often distorts morphological features critical to digital twin accuracy (e.g., P-wave duration or ST-segment slope). Advanced methods like adaptive filtering and Empirical Mode Decomposition (EMD) [6–8] improve adaptability to non-stationary noise but fail to preserve spectral integrity when noise overlaps with the diagnostic bandwidth of ECG. Such limitations introduce cascading errors in digital twin pipelines, where even minor waveform alterations can compromise predictive simulations of arrhythmias or drug responses.

In recent years, the application of deep learning methods to ECG signal denoising has achieved significant progress. For example, Autoencoders have been used for ECG denoising by reconstructing clean signals from noisy inputs [9, 16]. Deep Recurrent Neural Networks (RNNs) have been utilized for this purpose [10], and multi-branch Convolutional Neural Networks (CNNs), such as DeepFilter, have demonstrated excellent performance in removing specific types of noise. Despite these advancements, a fundamental challenge persists how to preserve local signal features while simultaneously capturing the global context. This difficulty arises because conventional CNNs or autoencoders [11, 16] typically operate with a limited receptive field, which restricts their ability to capture long-range dependencies. Conversely, Transformer-based models, [13, 15] excel at global modeling but may lose critical local details if they lack multi-scale feature pathways. Furthermore, neural networks [14, 17] often overlook the preservation and reconstruction of spectral information [18], which can lead to distortion in the recovered ECG signal. The classic TransUNet architecture incurs high computational cost at each CNN encoder and decoder stage.

Motivated by these challenges, we have designed a novel and lightweight TF-TransUNet1D, a lightweight yet powerful architecture specifically designed to meet the stringent signal fidelity requirements of cardiac digital twin systems. Our novel hybrid architecture combines the complementary strengths of local feature representation and global context aggregation through two key innovations: (1) a time-frequency constrained Transformer U-Net that simultaneously preserves both morphological details and spectral characteristics, and (2) a multi-domain loss function that jointly optimizes temporal waveform accuracy and spectral fidelity. This dual-domain optimization ensures clinically-relevant features (e.g., ST-segment morphology and P-wave characteristics) remain intact during denoising, which is crucial for reliable cardiac digital twins.

2 Methodology

2.1 Temporal-Aware Encoder-Decoder with Skip Connections

The encoder progressively compresses the input ECG feature representations. At the deepest level, after the D4 convolutional block, the feature map is fed into the Transformer module. Positional encoding is added to preserve temporal order, and the sequence is processed by N stacked Transformer encoder layers, each consisting of multi-head self-attention and feedforward sub-layers. The enhanced feature sequence is then reshaped (if necessary) back into spatial layout and passed to the U-Net decoder.

The decoder consists of Upsampling layers (transposed convolutions) U1–U4, each followed by a double convolution block. Skip connections from corresponding encoder stages are concatenated with the Upsampled feature maps (e.g., U1 output is concatenated with features from D3), helping to retain high-frequency ECG details that may have been lost during compression. The final layer is a 1D convolution with kernel size 1 (OutConv), which maps the decoded high-dimensional features back to the target output channels—1 for single-lead ECG denoising. TF-TransUNet1D takes input segments of shape (batch size, 1, 3600) and outputs denoised ECG segments of the same shape. All convolution layers use appropriate padding to maintain alignment and apply batch normalization to aid convergence (unless otherwise specified) (Fig. 1).

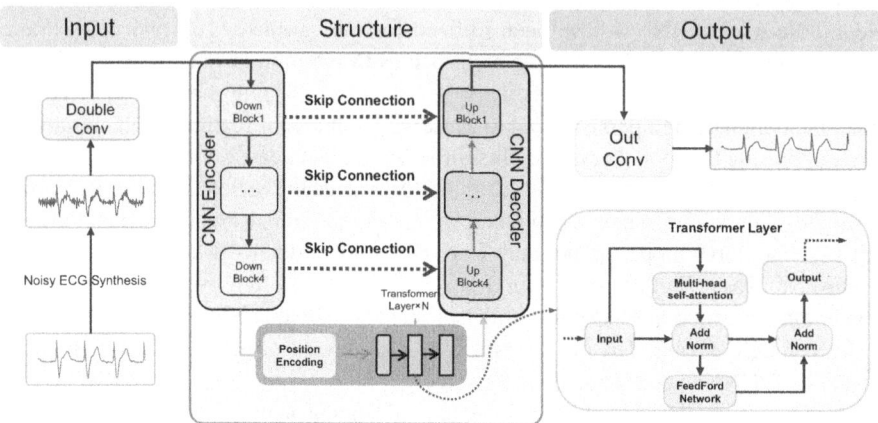

Fig. 1. A TF-TransUNet1D architecture, which follows a U-Net-style encoder-decoder design with four DownSampling/Upsampling levels and a Transformer encoder at the bottleneck. The encoder includes an input convolution block (INC) and three down sampling stages (D1–D4), each with two 1D convolutional layers (ReLU) and a max-pooling layer.

2.2 Modeling Long-Range Dependencies via Transformer

The Transformer encoder integrated into TF-TransUNet1D is designed to model long-range dependencies in ECG sequences. Each encoder layer follows the standard architecture: Multi-Head Self-Attention (MHSA) followed by a position-wise feedforward network. Formally, given an input sequence of feature vectors $X \in \mathbb{R}^{T \times d}$, where T is the sequence length and d is the feature dimension per token, the self-attention mechanism first computes query, key, and value matrices via learned linear projections of X. The self-attention output for a single head is computed as shown in Eq. (1), where a context-aware representation is produced by taking the weighted average of value vectors, with weights determined by the similarity between queries and keys. In the multi-head setting, attention is computed in parallel across H heads, and the results are concatenated and linearly transformed back to d-dimensions.

Residual skip connections and layer normalization (LN) are applied around both the MHSA and feedforward sub-layers. At the heart of the TransUNet-1D architecture, a standard Transformer encoder is embedded within the bottleneck of the U-Net. This design strategically combines the strong local feature extraction of CNN with the Transformer's ability to model global context. The U-Net encoder converts the raw ECG signal (length L) into a compact feature sequence of length $L' \ll L$ and channel size $c' \gg 1$. While rich in semantic information, this sequence's receptive field remains limited by the depth of CNN layers.

To overcome this, the Transformer encoder takes the compressed sequence as input and applies Multi-Head Self-Attention (MHSA) to model long-range dependencies. Each time step attends to all others, enabling the model to learn global temporal relationships—crucial for understanding the dynamics between QRS complexes, P waves, and T waves across the full 10-s segment. This global context helps distinguish physiological signals from non-stationary noise such as muscle artifacts. Through iterative refinement, the Transformer enhances meaningful patterns while suppressing irrelevant noise, yielding a purified feature representation for the decoder to reconstruct high-quality ECG signals.

2.3 Dual-Domain Loss Guided Strategy

To effectively guide the training of the TransUNet-1D model, we designed and adopted a multi-objective loss function that jointly optimizes both time-domain waveform reconstruction accuracy and frequency-domain spectral consistency. The total loss L_{loss} is a weighted sum of these two components. Let $y[n]$ denote the clean ECG signal and $\hat{y}[n]$ the denoised output from the model (for $n = 1, \ldots, N$ samples within a segment).

To balance convergence stability and robustness to artifacts, we ultimately chose the Smooth L1 loss. This loss behaves like mean squared error (MSE) when the absolute error $|y - \hat{y}| < \beta$, providing smooth gradients for precise convergence, and like mean absolute error (MAE) when $|y - \hat{y}| \geq \beta$, applying a linear penalty to reduce the influence of outliers and prevent overreactions to noise. This design allows the model to accurately reconstruct the underlying ECG morphology while remaining resilient to residual noise and prediction errors, resulting in more natural and reliable denoised signals.

In addition to time-domain fidelity, we introduce a spectral loss L_{spectral} to ensure that the denoised signal retains the correct frequency content compared to the ground truth. We compute the Fast Fourier Transform (FFT) of both signals and compare their magnitudes:

$$\mathcal{L}_{spectral} = \frac{1}{K} \sum_{k=1}^{K} \left(|Y[k]| - |\hat{Y}[k]| \right)^2, \tag{1}$$

where $Y[k]$ and $\hat{Y}[k]$ are the discrete Fourier magnitudes of the clean and reconstructed signals at frequency bin k, and K is the total number of frequency bins.

This term penalizes the discrepancy between the power spectra of $\hat{y}$ and y, guiding the model to not only minimize point-wise errors but also correctly redistribute energy across frequencies. This is crucial for preserving characteristic ECG bandwidths and preventing distortion of features such as QRS frequency components. The total training loss is a weighted sum of the two components:

$$\mathcal{L}_{total} = w_{time} \mathcal{L}_{time} + w_{spectral} \mathcal{L}_{spectral} \tag{2}$$

By optimizing L_{total}, TF-TransUNet1D learns to produce outputs that closely match the clean ECG in both the time and frequency domains. Incorporating L_{spectral} is especially beneficial under heavy noise conditions, as it prevents the network from simply smoothing the signal, a strategy that may minimize MSE but also attenuate important high-frequency components. Instead, the model is encouraged to recover the true frequency characteristics of the signal, resulting in denoising that is more physiologically accurate.

3 Experiments

3.1 Experiments Materials

Noisy Signal Synthesis and Preprocessing. Clean ECG signals from the MIT-BIH Arrhythmia Database are mixed with various noise types from the Noise Stress Test Database (NSTDB) to generate paired clean/noisy ECG segments with specific signal-to-noise ratios (SNRs). A sliding window mechanism is employed to divide the recordings into multiple shorter, fixed-length segments (each input segment comprises 3,600 sampling points, corresponding to a duration of 10 s) for model training and evaluation. Noise sources include baseline wander, muscle (EMG) noise, electrode motion artifacts, and powerline interference. All segments are z-normalized (zero mean, unit variance), and the same random noise instance is added to the corresponding clean segment to form paired training data. A portion of the dataset is reserved for validation to tune hyperparameters (e.g., loss weights, learning rate) and implement early stopping. The test set consists of ECG segments from subjects not seen during training, ensuring a fair assessment of the model's generalization ability.

Implementation. The deep learning framework was implemented from scratch using the PyTorch library, with custom modules defining the model architecture (model.py), data processing pipelines, and training procedures. The training process utilized AdamW with cosine annealing decay ($T\_max = 100, \eta\_min = 1e-6$) for optimization. All models were trained for 100 epochs with a batch size of 16 on NVIDIA GeForce RTX 4090GPU.

Evaluation Metrics. The performance of 1D ECG signal denoising is evaluated using standard metrics that assess both reconstruction fidelity and noise suppression. Signal-to-Noise Ratio (SNR) measures how effectively noise is reduced—higher values indicate better preservation of the original ECG waveform. SNR Improvement (SNRI), the difference between output and input SNR, reflects the model's noise reduction capability. Percentage Root Mean Square Difference (PRD) quantifies the normalized reconstruction error, with lower values indicating better retention of the signal's power. Pearson Correlation Coefficient (PCC) evaluates morphological similarity, showing how well the ECG waveform shape is preserved. Mean Absolute Error (MAE) measures the average amplitude difference between clean and denoised signals, reflecting point-wise reconstruction accuracy. Together, SNRI reflects noise suppression, PCC captures waveform integrity, and MAE assesses numerical precision—providing a comprehensive evaluation of ECG denoising performance.

3.2 Results

Quantitative Results Analysis. To evaluate the performance of the model, we conducted experiments using our synthesized noisy dataset on the proposed model as well as several commonly used baseline models, under the same settings such as initial learning rate, optimizer, and other parameters.

Table 1 presents a comparison of denoising performance using different models under the condition of mixed noise with a target SNR of 0 dB. Our proposed TF-TransUNet1D model, constrained by a time-frequency hybrid loss function, achieves the best results across all evaluation metrics, including MAE, PCC, and SNRI. In particular, in terms of Mean Absolute Error (MAE), our model reduces the error by approximately 24.3% compared to the second-best performing model, indicating that the average amplitude difference between the reconstructed and clean signals is minimal. Furthermore, combined with the high Pearson Correlation Coefficient, the results highlight the strong performance of the TF-TransUNet1D model guided by Dual-Domain Loss in preserving signal fidelity and spectral characteristics.

Table 1. Model Comparison for Denoising Tasks with Three-Type Signal Mixtures at 0 dB

Model	MAE	PCC	SNRI
CNN-LSTM	0.1819	0.916	10.61
U-Net 1D	0.1697	0.9195	10.85
FastRNN	0.2057	0.9138	9.74
TF-TransUNet1D	0.1285	0.9540	13.36

Figure 2 presents systematic experimental results on a synthesized noisy dataset, demonstrating the model's strong robustness and generalization when handling diverse single and combined noise types. Under the extreme low Signal-to-Noise Ratio (SNR = 0 dB), the model achieved the lowest Mean Absolute Error (MAE) of 0.0819 against Motion Artifact (MA) noise and the highest Pearson Correlation Coefficient (PCC) of 0.981 against Burst Noise (BW). When the SNR increased to 10 dB, the PCC further improved to 0.992, indicating high waveform reconstruction quality under mixed-frequency interference. While slight performance degradation occurred with composite noise (BW + Electromagnetic Interference (EM) + MA), the model's MAE and PCC remained within acceptable ranges. It consistently outperformed in Signal-to-Noise Ratio Improvement (SNRI), affirming the effectiveness of the dual-branch architecture and hybrid loss design in complex noise scenarios.

Qualitative Analysis of Results. The qualitative visualization of ECG denoising using TF-TransUNet1D (Fig. 3) demonstrates marked improvements in waveform morphology preservation and noise suppression. The model effectively smooths high-frequency myoelectric artifacts and attenuates substantial motion artifacts, restoring normal cardiac rhythm while maintaining diagnostically critical features—distinct P-waves, sharply defined QRS complexes, and well-preserved T-waves—with exceptional morphological, amplitude, and phase fidelity to gold-standard references. This success is largely

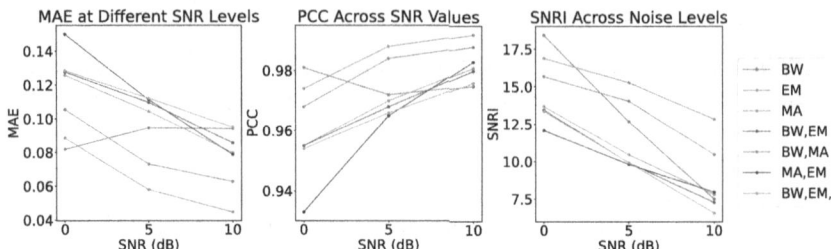

Fig. 2. Three-line plots depicting model performance across different noise types and target SNR levels. The figure consists of three subplots, each showing one evaluation metric (MAE, PCC, and SNRI) as a function of SNR (0 dB, 5 dB, and 10 dB). Each line corresponds to a specific noise type or noise combination, with a shared legend positioned on the right.

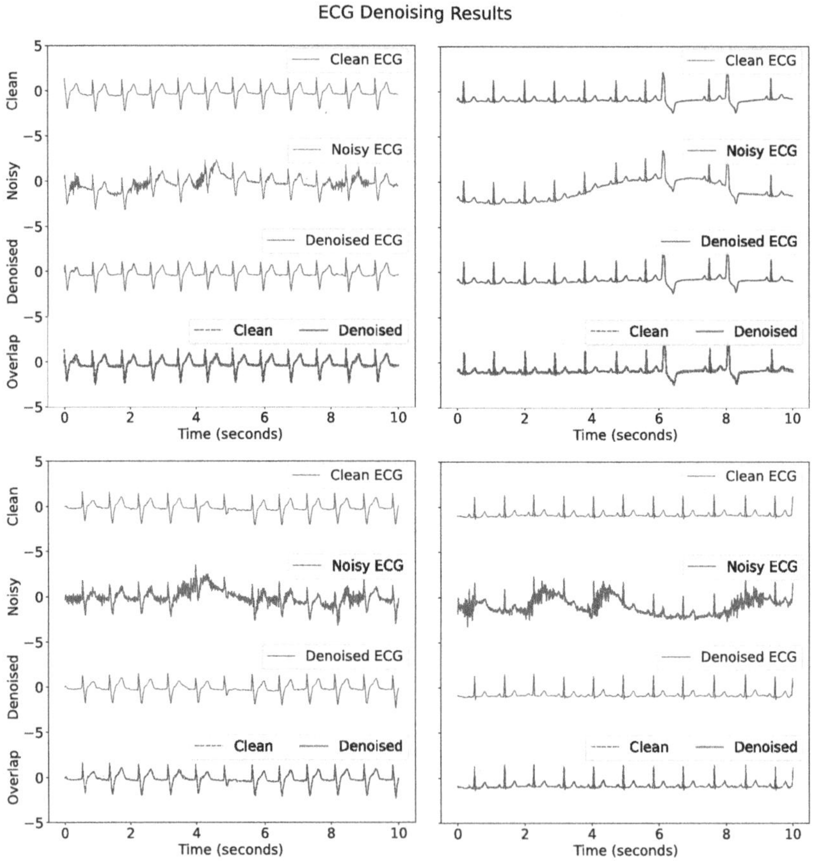

Fig. 3. Denoising Results of ECG Signals Using the TF-TransUNet1D Model

attributed to the incorporation of a frequency-domain loss component, which overcomes the limitations of traditional time-domain methods by ensuring spectral consistency, particularly within powerline interference bands, thereby preserving the integrity of clinically important waveforms. Comprehensive quantitative and qualitative evaluations confirm that TF-TransUNet1D delivers superior denoising performance and robust generalization for ECG signals. Furthermore, its lightweight design facilitates real-time deployment in clinical applications such as ambulatory monitoring and wearable medical devices.

4 Conclusion

In this work, we present TF-TransUNet1D, a Transformer-enhanced U-Net model designed for denoising 1D ECG signals. By integrating global self-attention with local feature extraction and skip connections, the architecture effectively bridges convolution- and attention-based approaches. It achieves robust noise reduction across various artifact types and noise levels, while preserving the morphological integrity of ECG waveforms. Driven by a multi-objective loss function that balances time-domain precision with frequency-domain fidelity, TF-TransUNet1D delivers denoised outputs that are both quantitatively accurate and qualitatively faithful. Importantly, this work offers an efficient and scalable strategy for incorporating Transformer blocks into biomedical time-series models, with minimal architectural overhead. By enhancing ECG signal quality, TF-TransUNet1D supports more reliable feature extraction and physiological modeling, making it a valuable component for building high-fidelity cardiac digital twins. The proposed framework contributes to the broader vision of personalized cardiovascular care, where accurate, artifact-free ECG data forms the foundation for dynamic patient-specific digital twin simulations.

Acknowledgment. This work was supported by NUS start-up funding to L. Li. We also thank the PhysioNet team for providing access to the MIT-BIH Arrhythmia Database.

References

1. Li, L., et al.: Personalized topology-informed localization of standard 12-lead ECG electrode placement from incomplete cardiac MRIs for efficient cardiac digital twins. Med. Image Anal. **101**, 103472 (2025)
2. Li, L., et al.: Toward enabling cardiac digital twins of myocardial infarction using deep computational models for inverse inference. IEEE Trans. Med. Imaging **43**(7), 2466–2478 (2024)
3. Kumar, A., Sharma, A., Pachori, R.B.: Stationary wavelet transform based ECG signal denoising method. ISA Trans. **114**, 251–262 (2021)
4. Romero, I., Geng, D., Berset, T.: Adaptive filtering in ECG denoising: a comparative study. In: Proceedings of the 2012 Computing in Cardiology Conference (2012)
5. Kabir, M.A., Shahnaz, C.: Denoising of ECG signals based on noise reduction algorithms in EMD and wavelet domains. Biomed. Signal Process. Control **7**(5), 481–489 (2012)

6. Sraitih, M., Jabrane, Y.: A denoising performance comparison based on ECG signal decomposition and local means filtering. Biomed. Signal Process. Control **69**, 102903 (2021)
7. Weng, B., Blanco-Velasco, M., Barner, K.E.: ECG denoising based on the empirical mode decomposition. In: Proceedings of the 2006 International Conference of the IEEE Engineering in Medicine and Biology Society
8. Li, C., Zhang, X., Wang, Y., Liu, H., Chen, Z., Liu, J.: ECG denoising method based on an improved VMD algorithm. IEEE Sens. J. **22**(23), 22725–22733 (2022)
9. Chiang, H.-T., Hsieh, Y.-Y., Fu, S.-W., Hung, K.-H., Tsao, Y., Chien, S.-Y.: Noise reduction in ECG signals using fully convolutional denoising autoencoders. IEEE Access **7**, 60806–60813 (2019)
10. Antczak, K.: Deep recurrent neural networks for ECG signal denoising. arXiv preprint arXiv: 1807.11551 (2018)
11. Dasan, E., Panneerselvam, I.: A novel dimensionality reduction approach for ECG signal via convolutional denoising autoencoder with LSTM. Biomed. Signal Process. Control **63**, 102225 (2021)
12. Li, H., Zhou, Y., Liu, H., Peng, Y., Zhang, J., Li, Y., Wang, Y.: Descod-ECG: deep score-based diffusion model for ECG baseline wander and noise removal. IEEE J. Biomed. Health Inf. (2023)
13. Singh, P., Sharma, A.: Attention-based convolutional denoising autoencoder for two-lead ECG denoising and arrhythmia classification. IEEE Trans. Instrum. Measur. **71**, 1–10 (2022). Art. no. 4007710
14. Hou, Y., Liu, R., Shu, M., Xie, X., Chen, C.: Deep neural network denoising model based on sparse representation algorithm for ECG signal. IEEE Trans. Instrum. Measur. **72**, 1–11 (2023). Art. no. 2507711
15. Zhu, D., Chhabra, V.K., Khalili, M.M.: ECG Signal Denoising Using Multi-scale Patch Embedding and Transformers. arXiv preprint arXiv:2407.11065 (2024)
16. Lin, H., Liu, R., Liu, Z.: ECG signal denoising method based on disentangled autoencoder. Electronics **12**(7), 1606 (2023)
17. Qiu, L., et al.: Two-stage ECG signal denoising based on deep convolutional network. Physiol. Meas. **42**(11), 115002 (2021)
18. Mourad, N.: ECG denoising algorithm based on group sparsity and singular spectrum analysis. Biomed. Signal Process. Control **50**, 62–71 (2019)

DeformMLP: Effective Deformation Prediction for Breast Cancer Using Graph Topology-Assisted MLPs

Yong-Min Shin[1], Kyunghyun Lee[1], Sunghwan Lim[2], Kyungho Yoon[1(✉)], and Won-Yong Shin[1(✉)]

[1] Yonsei University, Seoul, South Korea
{yoonkh,wy.shin}@yonsei.ac.kr
[2] Korea Institute of Science and Technology, Seoul, South Korea

Abstract. Early diagnosis of breast cancer is crucial, enabling the establishment of appropriate treatment plans and markedly enhancing patient prognosis. While direct magnetic resonance imaging (MRI)-guided biopsy demonstrates promising performance in detecting cancer lesions, its practical application is limited by prolonged procedure times and high costs. To overcome these issues, an indirect MRI-guided biopsy that allows the procedure to be performed outside of the MRI room has been proposed, but it still faces challenges in creating an accurate *real-time* deformable breast model. In our study, we propose DEFORMMLP, a deformation prediction method that uses *graph topology*-assisted multilayer perceptrons (MLPs) as the main backbone architecture. DEFORMMLP is able to effectively predict the deformation of nodal surfaces given a point force with significantly faster training and low memory requirements. As DEFORMMLP is designed to take force vectors and graph features as input, along with nontrivial *graph structure encoding*, which performs feature propagation based on the underlying graph constructed from the element information. Our experimental results demonstrate that DEFORMMLP outperforms graph neural network (GNN)-based alternatives with respect to both test root mean squared error (RMSE) and efficiency in time and memory costs. The source code is publicly available at https://github.com/jordan7186/DeformMLP.

Keywords: Breast cancer · deformation prediction · graph neural network · graph topology · multilayer perceptron

1 Introduction

Breast cancer is the most prevalent cancer globally and the leading cause of cancer-related deaths among women [3]. Early diagnosis enables the establishment of appropriate treatment plans and improves prognosis, making it a critical

---
K. Yoon and W. Y. Shin contributed equally as senior authors.

© The Author(s), under exclusive license to Springer Nature Switzerland AG 2026
L. Li et al. (Eds.): DT4H 2025, LNCS 16193, pp. 99–108, 2026.
https://doi.org/10.1007/978-3-032-07694-6_10

component of breast cancer management [24]. The most common and preferred method for accurate diagnosis of breast cancer is a core needle biopsy, which is typically guided by mammograms and ultrasound imaging [6].

Meanwhile, breast magnetic resonance imaging (MRI) is recognized for its superior sensitivity and specificity in detecting suspicious cancer lesions [16]. Research indicates that more than half of these lesions are visible only on breast MRI [2]. To leverage this advantage, direct MRI-guided breast biopsy, which involves performing the biopsy in the MRI suite, has been validated in clinical settings [19]. However, this procedure is currently limited to a small number of women due to constraints such as lengthy procedure times, high costs, and difficulties with biopsy positioning [22].

To address this, an indirect MRI-guided breast biopsy utilizing machine vision techniques, which involves performing biopsy outside the MRI suite, has been proposed in [10]. This method involves estimating the positions of MRI targets within the breast through real-time deformable registration between a deformable breast model and real-time shape sensing data of the breast. Despite this study, creating a *real-time* deformable breast model that is both responsive and accurate still remains a critical challenge.

Recently, training machine learning (ML) models for effective prediction of physical phenomena has gained attention as a strong candidate for achieving both responsiveness and accurate performance [5,15,23,32]. Although ML-based approaches can utilize the abundance of the software tools to train ML models alongside their general applicability to learn from data (resulting in several medical applications), sufficient curation and generation of (training) data acts as a key aspect in the overall process. While a finite element (FE) mesh is typically used to model the tissue/organ of interest, lots of studies have employed graph neural networks (GNNs) as their main architecture to learn from the mesh data due to their high ability to naturally encode both the nodal features (*e.g.*, force vectors) as well as the structural information from the mesh [7,11,18,21]. Nonetheless, using GNNs is often known to introduce several inherent limitations, such as slow inference speed and potentially memory-intensive training.

In our study, we tackle the problem of deformation prediction of breast models by building an efficient and accurate ML-based method under different force vectors. As our first main contribution, we curate a *mesh-based breast tissue deformation* dataset by pre-processing MRI images from a breast phantom. Subsequently, FE analysis is used to simulate the deformation under diverse force direction applied to different places on the mesh. Then, a straightforward approach would be to employ a GNN to leverage the mesh information, which however brings inherent limitations. For example, it is common to stack up to 2–3 GNN layers due to oversmoothing (*i.e.*, performance degradation for deeper layers [4,14,17]). Furthermore, GNNs are also known to having exponentially slow inference and training speed with deeper models [8,29–31]. To solve this problem, as our second main contribution, we propose DEFORMMLP, an effective alternative of deformation prediction built upon multilayer perceptrons (MLPs). DEFORMMLP utilizes both the force vectors and *graph structure encoding vectors* to inject the graph topology information to the MLP. More specifically, to encode the graph structural information, we initialize the fixed nodes with positional encoding and perform feature propagation over the graph. This allows global information to be used by the model without oversmoothing. Our extensive experiments demonstrate DEFORMMLP can achieve better performance in test root mean square error (RMSE), while requiring fewer memory with faster training time compared to GNNs.

2 Dataset Generation via FE Analysis

Unlike the case where datasets are publicly available, the training dataset in our study is obtained from *solutions of FE analysis* as shown in Fig. 1. In this section, we elaborate on both the proposed computational FE model, used to generate physics information on how breast tissue deforms, and the method used for training data generation.

2.1 FE Formulation

The tissue deformation is governed by the continuum mechanics based local equilibrium equations [1], as given by $\tau_{ij,j} + f_i^B = 0$ in Ω, where $\tau_{ij,j}$ is the derivative of the ij component of the stress tensor with respect to j, f_i^B is the i-directional component of the applied body force, and Ω represents the volume region to be analyzed.

According to the standard procedure in [28], the FE formulation to be solved for soft tissue deformation analysis is obtained as

$$\mathbf{KU} = \mathbf{F}, \tag{1}$$

where $\mathbf{K}$ is a global stiffness matrix of the given structural system, $\mathbf{U}$ is a vector representing the displacement at the FE nodes, and $\mathbf{F}$ is an external force vector applied at the nodes.

2.2 FE Modeling Using MR Images

A breast phantom containing an internal tumor mass (named breast model 3401, GPI Anatomicals, IL) is prepared. To obtain the structural information of the phantom, a T_1-weighted high-resolution image (field of view $15 \times 15 \, \text{cm}^2$, slice thickness 1.0 mm, image matrix 160×160, number of slices 150, repetition time 6.193 ms, echo time 2.303 ms, flip angle $10°$) is obtained by the 3 T magnetic resonance (MR) scanner (Achieva, Philips, Netherlands).

The breast tissue, due to its significantly higher signal intensity in T_1-weighted images compared to the background signal, is segmented by applying a simple threshold to obtain 3D geometric information, which is exported in the OBJ file format. Subsequently, the obtained OBJ file is used to create 3D tetrahedral meshes using the ANSYS SpaceClaim tool (named SpaceClaim 2021

R1). The target quality of the mesh is set to 0.05, resulting in an average surface area of $2.6473 \times 10^{-5} m^2$ and a minimum edge length of $2.2894 \times 10^{-3} m$. These parameters are confirmed to provide sufficient accuracy through solution convergence testing. 10-node tetrahedral quadratic elements (named Tet10 in ANSYS) are assigned to the meshes for constructing the FE model, resulting in 7,598 elements, 13,311 nodes, and 39,933 degrees of freedom. Using the connectivity information of the nodes consisting the FEs, the local stiffness matrix of each tetrahedral element is assembled to obtain the global stiffness matrix **K** in Eq. (1).

The degrees of freedom at the nodes on the bottom surface are fixed and used as boundary conditions (BCs). A linear elastic material law is applied to all the elements, being characterized by a Young's modulus of 690 Pa and a Poisson's ratio of 0.495.

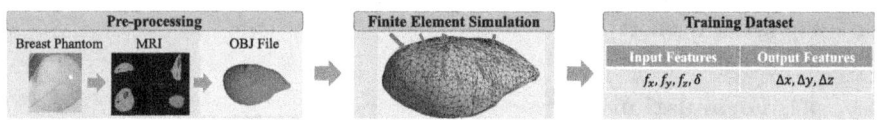

Fig. 1. The schematic overview of training data generation.

2.3 Training Data Generation

As our training dataset, physics information on the breast deformation under given surface force condition is generated using the proposed FE model. The resulting dataset is acquired for 300 different force directions and magnitudes at ten force locations, resulting in a total of 1,000 input force vectors. To consider only situations where compressive deformation occurs, the direction of each force is randomly assigned with the constraint that the vertical component (i.e., the y-direction) is set to be negative. The magnitude of the force, denoted as M, is also randomly given by $M = 0.02 \times (k + 0.5)$, where $k \in [0, 1]$ is drawn from a uniform distribution. As a result of the FE analysis, 1,000 output displacement vectors corresponding to each input force vector are obtained. All the simulations are carried out using the commercial FE analysis software ANSYS (named Mechanical Products 2021 R1).

Based on these input force vectors of the FE model, the x-, y-, and z-directional forces (denoted as f_x, f_y, and f_z, respectively) at all the FE nodes are selected as input features for DEFORMMLP. Additionally, we include the displacement BCs (δ) as an input feature where $\delta = 0$ represents fixed BCs and $\delta = 1$ corresponds to free conditions. As output features, the x-, y-, and z-directional displacements (denoted as Δx, Δy, and Δz, respectively) obtained from the output displacement vectors of the FE analysis are used.

3 Proposed Method: DeformMLP

In this section, we first describe our graph structure encoding method, which is used to design DEFORMMLP as input representing the *proximity context* while predicting the deformation. Then, we elaborate on the architecture design of DEFORMMLP.

3.1 Graph Construction

In order to perform deformation prediction, our method is required to utilize two types of information: 1) the force vectors and 2) the mesh structure of the target node from the force node, which is then represented as a graph. In our experiments, we construct a graph based on the 10-node tetrahedral quadratic element information: **(Step 1)** We first connect all nodes that belong to each element. **(Step 2)** From the dense graph, we eliminate edges that appear only in a single element.

3.2 Graph Structure Encoding

GNN models are advantageous when graph topological information are beneficial due to their expressive capability through the message passing mechanism, eventually producing node representations based on both the node features and the graph structure. In DEFORMMLP, we do not employ any message passing as part of the architecture and only use an MLP. This motivates us to device a new two-stage pre-processing step that encode the graph structure as node features, which will be fed into the MLP as input.

(Step 1) Only for the fixed nodes, their node feature vectors $\mathbf{v} \in \mathbb{R}^3$ containing the 3D information are transformed by *positional encoding* [25,26] as $[\cdots, \cos(2\pi\sigma^{(j/m)}\mathbf{v}), \sin(2\pi\sigma^{(j/m)}\mathbf{v}), \cdots]$, where $j \in \{0, \cdots, m-1\}$, $\sigma = 0.1$, and $m = 6$. The remaining nodes are initialized as zero vectors.

(Step 2) We perform T steps of *feature propagation* [20,33] over the whole graph. We denote the node feature matrix in the t-th iteration as $\mathbf{X}^{(t)}$ (note that $\mathbf{X}^{(0)}$ represents the initial features in Step 1), the adjacency matrix of the graph as $\mathbf{A}$, the diagonal degree matrix as $\mathbf{D}$, the mask matrix indicating nodes at BCs as $\mathbf{M} \in \{0,1\}^{N \times 6m}$ (whose entry is 1 for BCs). Then, the feature propagation is formalized as follows:

$$\mathbf{X}'^{(t+1)} = \mathbf{D}^{-1/2}\mathbf{A}\mathbf{D}^{-1/2}\mathbf{X}^{(t)} \qquad (2)$$

$$\mathbf{X}^{(t+1)} = (\mathbf{1}_{N \times 6m} - \mathbf{M}) \odot \mathbf{X}'^{(t+1)} + \mathbf{M} \odot \mathbf{X}^{(0)}, \qquad (3)$$

where $\odot$ is the element-wise product and $\mathbf{1}_{N \times 6m}$ denotes the all-one matrix with the dimension of $N \times 6m$. We use $T = 6$, which covers the whole graph.

The graph structure encoding can be interpreted as solving a heat diffusion equation along the graph structure with a constant heat source from the BC nodes [20], where the heat source contains the 3D information in our case.

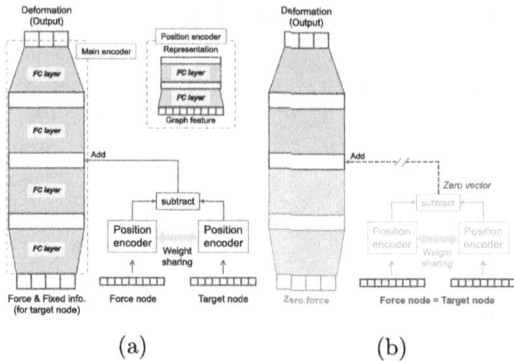

Fig. 2. The schematic overview of DEFORMMLP. (Color figure online)

3.3 Architecture Design

We aim to develop our deformation prediction model according to the following two objectives. Specifically, for *efficiency*, we build a neural network model that does not involve graph convolution layers and only use an MLP as our backbone architecture. Also, for *accuracy*, the model jointly incorporate the force vectors as well as the graph feature as input.

To this end, we propose DEFORMMLP, based on fully connected layers (without biases and ReLU activation) that operate on individual nodes and predicts its deformation (*i.e.*, Δx, Δy, and Δz). Specifically, for a given target node, DEFORMMLP takes (i) the force vector (along with whether the target node is fixed), (ii) the force node, and (iii) the target node as input, and predicts the deformation as output. The input (i) is fed into a 4-layer *main encoder* MLP, while a separate 2-layer *position encoder* MLP produces the vector representations (*i.e.*, the graph structure encoded-vectors) for the input (ii) and (iii), which are subtracted with each other before being added to the intermediate hidden representation of the input (i) (see Fig. 2a). By this design, DEFORMMLP is able to effectively discriminate between two different scenarios. **(Case 1)** When the target node is the force node, the position encoder eventually produces a zero vector after subtraction (see the teal part in Fig. 2b), and the model predicts the deformation based on the input force. **(Case 2)** When the force vector is zero, the first two layers of the main encoder (see the orange part in Fig. 2b) are naturally deactivated, and the model predicts the deformation based on the encoded graph structural information. Such a design allows DEFORMMLP to use its learnable parameter effectively during training, while making full use of the graph topological information without any message passing.

4 Experimental Results and Analyses

In this section, we describe the basic experimental settings. Then, we carry out extensive experiments and analyze their results.

4.1 Experimental Settings

The 1,000 data samples generated from Sect. 2 are randomly split into training, validation, and test sets with a ratio of 8:1:1. The validation set is used for hyperparameter tuning and early stopping, where the patience is set to 10 with the maximum epoch of 256. As baseline, we employ two 2-layer GNN models, using GCN [13] and GraphSAGE [9]. The node features for the GNN models are generated by concatenating the force and 3D position vectors for fair comparison. We use the Adam optimizer [12] alongside the mean square loss.

4.2 Test RMSE Results and Analyses

We evaluate the performance accuracy and analyze its results. Table 1 compares the performance of deformation prediction for DEFORMMLP against GCN [13] and GraphSAGE [9] with respect to the test RMSE for the test dataset. We measure the test RMSE on average across all nodes in the graph. Furthermore, we analyze the test RMSE with respect to three different groups: (1) the node to which the force is applied (*'force node'*), (2) nodes nearby the force node where there is no force but the output deformation is non-zero (*'nearby node'*), and (3) the remaining nodes (*'distant node'*). Our observations are made as follows:

- On average across all nodes, DEFORMMLP is the best performer, revealing a test RMSE of $0.1451mm$, which is apparently lower than that of both GCN and GraphSAGE.
- DEFORMMLP achieves the best performance on the nearby and distant nodes. In particular, DEFORMMLP exhibit extraordinarily better performance on the distance nodes up to 8 orders of magnitude compared to the benchmark methods.
- On the force nodes, the performance of DEFORMMLP is comparable to that of the benchmark methods.

Overall, we conclude that DEFORMMLP manifests superiority to a large extent in accurately predicting breast tissue deformations.

Table 1. Comparison of the test RMSE performance for different methods (in mm).

Method	Average	Force node	Nearby node	Distant node
GCN	0.1528	6.0008	0.2344	0.1798
GraphSAGE	0.1528	**5.9991**	0.1591	0.0543
DEFORMMLP	**0.1451**	6.0041	**0.1523**	$\mathbf{2.0290 \times 10^{-5}}$

4.3 Efficiency Results and Analyses

We analyze the efficiency of DEFORMMLP and the two GNN methods. We measure two aspects in efficiency: wall-time per training step (*computational complexity*) and peak GPU memory during training (*space complexity*).

Table 2 summarizes the empirical findings for time and memory requirements of three different methods. Firstly, we observe GNNs are noticeably more expensive in terms of time and memory. The message passing mechanism requires recursively aggregating information of neighboring nodes for every feed-forward process, which results in the time complexity to exponentially increase with respect to the number of layers [27]. Similarly, GNN-based approaches require full or batch loading of the underlying graph into memory [31]. We also observe DEFORMMLP achieves the fastest training time per step and requires the lowest memory consumption, as it does not employ message passing that is often expensive in terms of memory consumption.

Table 2. Efficiency comparison for different methods.

	DEFORMMLP	GCN	GraphSAGE
Time	**0.01061 s/step**	0.02438 s/step	0.01930 s/step
Memory	**872.54 MB**	2625.91 MB	1676.55 MB

Table 3. Performance comparison in test RMSE for different graph construction strategies (in mm).

Graph const.	Force node	Nearby node	Distant node
Step 1	**5.9329**	0.3283	0.3453
Step 1 + Step 2	6.0041	**0.1523**	$\mathbf{2.0290 \times 10^{-5}}$

4.4 Further Analysis: Ablation Study

We now conduct an ablation study with an alternative that constructs a dense graph by performing Step 1 only (with a removal of Step 2) in Sect. 3.1. Table 3 shows that the dense graph (the first row) results in slightly better performance on the force nodes, but significantly worse performance on the nearby and distant nodes. This is due to the fact that performing Step 2 indeed yields more informative graph features during feature propagation.

5 Conclusions

In this study, we tackled the problem of predicting breast tissue deformation in an end-to-end fashion. We deviced a deformable breast model by pre-processing MRI images from a breast phantom, and curated a dataset by simulating the deformation under different force vectors using FE analysis. Additionally, we proposed DEFORMMLP, an MLP-based method that is capable of effectively predicting the deformation by judiciously leveraging the force vectors and the mesh structure via graph structure encoding. Extensive experiments demonstrated (a) the efficiency of DEFORMMLP in training and memory costs as well as (b) the superiority of DEFORMMLP in the test RMSE performance, compared to GNN-based benchmark methods. Future research includes the extension of DEFORMMLP to more complex circumstances where the force is applied to a surface area rather than a single point.

Acknowledgments. This work was supported by the National Research Foundation of Korea (NRF) funded by Korea Government (MSIT) under Grant RS-2023-00220762.

Disclosure of Interests. All authors declare that they have no known competing financial interests or personal relationships that may influence the work reported in this paper.

References

1. Bathe, K.J.: Finite element procedures. Klaus-Jurgen Bathe (2006)
2. Berg, W.A., et al.: Detection of breast cancer with addition of annual screening ultrasound or a single screening MRI to mammography in women with elevated breast cancer risk. J. Am. Med. Assoc. **307**(13), 1394–1404 (2012)
3. Bray, F., Ferlay, J., Soerjomataram, I., Siegel, R.L., Torre, L.A., Jemal, A.: Global cancer statistics 2018: Globocan estimates of incidence and mortality worldwide for 36 cancers in 185 countries. CA Cancer J. Clin. **68**(6), 394–424 (2018)
4. Chen, D., Lin, Y., Li, W., Li, P., Zhou, J., Sun, X.: Measuring and relieving the over-smoothing problem for graph neural networks from the topological view. In: AAAI. New York, NY (2020)
5. Choi, M., Jang, M., Yoo, S.S., Noh, G., Yoon, K.: Deep neural network for navigation of a single-element transducer during transcranial focused ultrasound therapy: proof of concept. IEEE J. Biomed. Health Inform. **26**(11), 5653–5664 (2022)
6. Clinic, M.: Breast biopsy (2023). https://www.mayoclinic.org/tests-procedures/breast-biopsy/about/pac-20384812
7. Dalton, D., Gao, H., Husmeier, D.: Emulation of cardiac mechanics using graph neural networks. Comput. Methods Appl. Mech. Eng. **401**, 115645 (2022)
8. Guo, Z., Shiao, W., Zhang, S., Liu, Y., Chawla, N.V., Shah, N., Zhao, T.: Linkless link prediction via relational distillation. In: ICML. Honolulu, HI (2023)
9. Hamilton, W.L., Ying, Z., Leskovec, J.: Inductive representation learning on large graphs. In: NeruIPS. Long Beach, CA (2017)
10. Jeon, S., Kim, Y., Lim, S.: Out-bore MRI-guided breast biopsy employing machine vision and augmented reality techniques: preliminary study. In: EMBC. Orlando, FL (2024)
11. Jiang, C., Chen, N.Z.: Graph neural networks (GNNs) based accelerated numerical simulation. Eng. Appl. Artif. Intell. **123**, 106370 (2023)
12. Kingma, D.P., Ba, J.: Adam: a method for stochastic optimization. In: ICLR, San Diego, CA (2015)
13. Kipf, T.N., Welling, M.: Semi-supervised classification with graph convolutional networks. In: ICLR, Toulon, France (2017)
14. Li, Q., Han, Z., Wu, X.: Deeper insights into graph convolutional networks for semi-supervised learning. In: AAAI, New Orleans, LA (2018)
15. Liang, L., Liu, M., Martin, C., Sun, W.: A deep learning approach to estimate stress distribution: a fast and accurate surrogate of finite-element analysis. J. R. Soc. Interface **15**(138), 20170844 (2018)

16. Mann, R.M., Kuhl, C.K., Moy, L.: Contrast-enhanced MRI for breast cancer screening. J. Magn. Reson. **50**(2), 377–390 (2019)
17. Oono, K., Suzuki, T.: Graph neural networks exponentially lose expressive power for node classification. In: ICLR, Addis Ababa, Ethiopia (2020)
18. Pagan, D.C., Pash, C.R., Benson, A.R., Kasemer, M.P.: Graph neural network modeling of grain-scale anisotropic elastic behavior using simulated and measured microscale data. NPJ Comput. Mater. **8**(1), 259 (2022)
19. Price, E.R.: Magnetic resonance imaging-guided biopsy of the breast: fundamentals and finer points. Magn. Reson. Imaging Clin. N. Am. **21**(3), 571–581 (2013)
20. Rossi, E., Kenlay, H., Gorinova, M.I., Chamberlain, B.P., Dong, X., Bronstein, M.M.: On the unreasonable effectiveness of feature propagation in learning on graphs with missing node features. In: LoG, Virtual Event (2022)
21. Salehi, Y., Giannacopoulos, D.: PhysGNN: a physics-driven graph neural network based model for predicting soft tissue deformation in image-guided neurosurgery. In: NeurIPS, New Orleans, LA (2022)
22. Saslow, D., et al.: American cancer society guidelines for breast screening with MRI as an adjunct to mammography. CA Cancer J. Clin. **57**(2), 75–89 (2007)
23. Shin, M., Seo, M., Yoo, S.S., Yoon, K.: tfusformer: physics-guided super-resolution transformer for simulation of transcranial focused ultrasound propagation in brain stimulation. IEEE J. Biomed. Health Inform. **28**, 4024–4035 (2024)
24. Smith, R.A., et al.: American cancer society guidelines for breast cancer screening: update 2003. CA Cancer J. Clin. **53**(3), 141–169 (2003)
25. Tancik, M., et al.: Fourier features let networks learn high frequency functions in low dimensional domains. In: NeurIPS, Virtual Event (2020)
26. Vaswani, A., et al.: Attention is all you need. In: NeurIPS, Long Beach, CA (2017)
27. Yan, B., Wang, C., Guo, G., Lou, Y.: TinyGNN: learning efficient graph neural networks. In: KDD, Virtual Event (2020)
28. Yoon, K., Lee, P.S.: Nonlinear performance of continuum mechanics based beam elements focusing on large twisting behaviors. Comput. Methods Appl. Mech. Eng. **281**, 106–130 (2014)
29. Zeng, H., Zhou, H., Srivastava, A., Kannan, R., Prasanna, V.K.: GraphSAINT: graph sampling based inductive learning method. In: ICLR, Addis Ababa, Ethiopia (2020)
30. Zhang, D., et al.: AGL: a scalable system for industrial-purpose graph machine learning. Proc. VLDB Endow. **13**(12), 3125–3137 (2020)
31. Zhang, S., Liu, Y., Sun, Y., Shah, N.: Graph-less neural networks: teaching old MLPs new tricks via distillation. In: ICLR, Virtual Event (2022)
32. Zhao, W., Wang, W., Tian, Y.: Graformer: graph-oriented transformer for 3d pose estimation. In: CVPR, New Orleans, LA (2022)
33. Zhou, D., Bousquet, O., Lal, T.N., Weston, J., Schölkopf, B.: Learning with local and global consistency. In: NeurIPS, Vancouver and Whistler, Canada (2003)

Rule-based Key-Point Extraction for MR-Guided Biomechanical Digital Twins of the Spine

Robert Graf[1,2(✉)], Tanja Lerchl[1], Kati Nispel[1], Hendrik Möller[1,2], Matan Atad[1,2], Julian McGinnis[2,4], Julius Maria Watrinet[3], Johannes Paetzold[5], Daniel Rueckert[2,6], and Jan S. Kirschke[1]

[1] Department of Diagnostic and Interventional Neuroradiology, School of Medicine, TUM University Hospital, Technical University of Munich, Munich, Germany
robert.graf@tum.de
[2] Institut für KI und Informatik in der Medizin, TUM University Hospital, Technical University of Munich, Munich, Germany
[3] Sports Orthopedics Department, Klinikum Rechts der Isar, Technical University of Munich, Munich, Germany
[4] Department of Neurology, School of Medicine, Technical University of Munich, Munich, Germany
[5] Department of Radiology, Weill Cornell Medicine, New York, USA
[6] Department of Computing, Imperial College London, London, UK

Abstract. Digital twins offer a powerful framework for subject-specific simulation and clinical decision support, yet their development often hinges on accurate, individualized anatomical modeling. In this work, we present a rule-based approach for subpixel-accurate key-point extraction from MRI, adapted from prior CT-based methods. Our approach incorporates robust image alignment and vertebra-specific orientation estimation to generate anatomically meaningful landmarks that serve as boundary conditions and force application points, like muscle and ligament insertions in biomechanical models. These models enable the simulation of spinal mechanics considering the subject's individual anatomy, and thus support the development of tailored approaches in clinical diagnostics and treatment planning. By leveraging MR imaging, our method is radiation-free and well-suited for large-scale studies and use in underrepresented populations. This work contributes to the digital twin ecosystem by bridging the gap between precise medical image analysis with biomechanical simulation, and aligns with key themes in personalized modeling for healthcare.

Keywords: MRI Spine Landmark Extraction · Biomechanical Modeling · subject-specific

1 Introduction

Biomechanical modeling plays a critical role in understanding the mechanical behavior of the human spine and in studying musculoskeletal disorders. Finite element methods (FEM) are commonly used to assess local tissue stresses and deformations, supporting research into intervertebral disc (IVD) degeneration, scoliosis, and implant optimization [1,4,5,16]. Multibody systems (MBSs), on the other hand, capture the kinematics and dynamics of rigid body segments like vertebrae and are used to simulate spinal posture, joint loading, and musculoskeletal motion [3,14,20]. A core requirement for MBSs is the accurate identification of points of interest (POIs) on bones, which define joint axes, force application sites, and coordinate frames [10,13]. Lerchl et al. [13] introduced a rule-based framework for extracting such POIs from CT scans, achieving voxel precision of attachments of ligaments and muscles. This enabled subject-specific modeling of spinal mechanics. However, CT-based modeling is limited in determining soft tissues such as IVDs. To address this, we adapt and extend this method for use with MRI, which offers superior soft tissue contrast but poses challenges due to its lower resolution in sagittal T2-weighted sequences and anisotropic voxels, making traditional pixel-based algorithms less reliable. Moreover, this existing approach often assumes vertebrae are aligned with image axes, a problematic assumption in patients with spinal deformities like scoliosis. We overcome this by computing vertebral orientations directly from the image, enabling robust modeling even in rotated or misaligned scans. To promote reproducibility and further research, we release our implementation as an open-source script, facilitating MRI-based MBM workflows and expanding biomechanical digital twin applications to settings where soft tissue characterization is essential (Figs. 1 and 2).

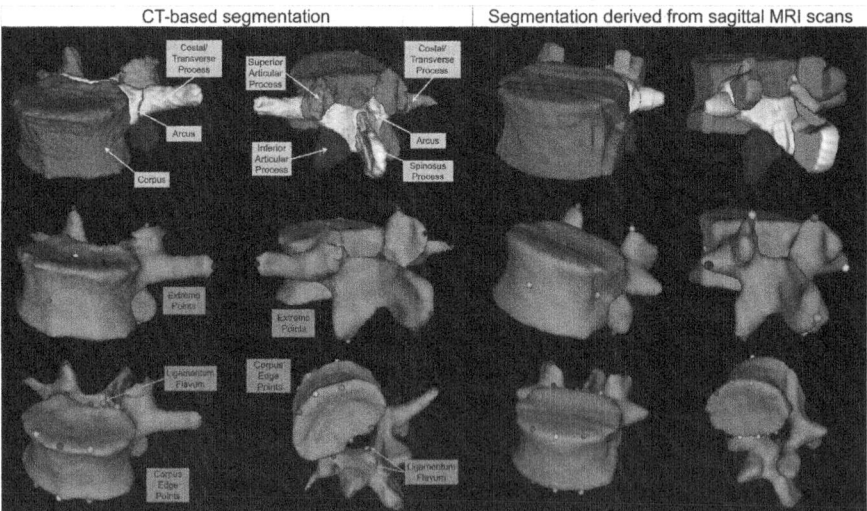

Fig. 1. Example of two lumbar vertebrae. The left example is derived from 1 mm isotropic CT, the right from sagittal MRI with a resolution of 3.3 mm in the left-right direction. Top row: Subregion of the vertebra used for analysis. Middle row: Extreme points. Bottom row: Corpus edge and ligamentum flavum points.

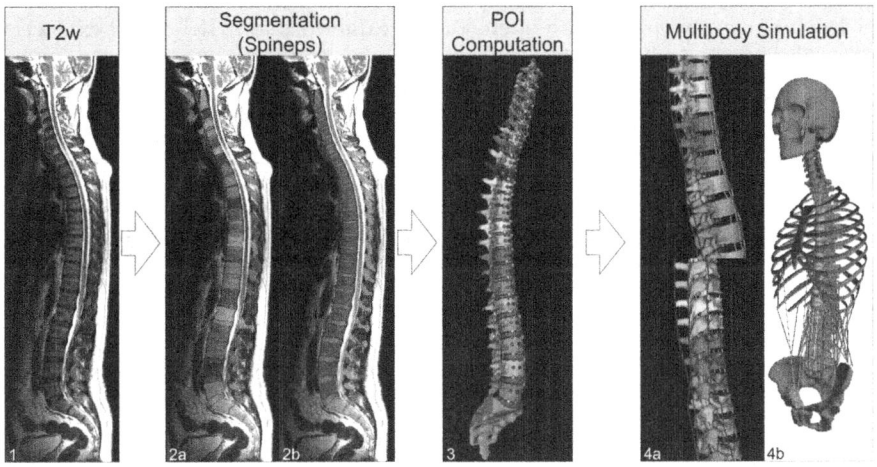

Fig. 2. Full pipeline for the generation of multibody models from MR imaging. A sagittal T2-weighted MRI (1) is segmented using Spineps, a machine learning based pipeline for automated whole spine segmentation of level-wise vertebrae and intervertebral discs (2a) as well as respective subregions (2b). Based on these segmentation masks, individual points of interest (POIs) are calculated (3) to define ligament (4a) and spinal muscle attachments to other regions (4b) in the subsequent multibody model of the torso.

Building on our approach to directly estimate vertebral orientation from image data, we situate our work within a broader landscape of anatomical landmark detection. Traditionally, POI prediction has focused on bone landmarks, such as those in the head [12,17,19] or lower limbs [6,7]. Multi-stage prediction networks are commonly employed to refine these estimates across successive processing steps. Several methods have been developed to estimate vertebral orientation, anatomical lines, or discrete landmarks. For instance, Galbusera et al. [8] used a ResNet-based regression model to directly estimate 3D vertebral axes from sagittal radiographs, achieving angular errors below 3° in more than 86% of cases. More recent graph-based strategies, such as Burgin et al. [2], detect pedicle and vertebral body landmarks using a U-Net, which are then processed via a graph neural network to infer vertebral pose and inter-point relationships. Other work has focused on line-based representations; Zhang et al. [21] proposed a dual-coordinate model that reconstructs vertebral lines from sparse landmark inputs, improving resilience to anatomical variation and partial visibility. In surgical planning contexts, Zhang et al. [22] introduced a YOLO-inspired network that jointly regresses vertebral translation and orientation as quaternions, achieving angular errors around 2.55°. Lastly, landmark-based methods like that of Khanal et al. [11] use vertebral corner regression to estimate vertebral tilt angles, which are especially relevant for scoliosis analysis. However, despite this progress, there remains no publicly available benchmark for vertebral orientation estimation, and most current approaches rely heavily on labor-intensive manual annotations.

In summary, we present an open-source framework for MRI-based extraction of vertebral orientation and points of interest, enabling more accurate multi-body spinal modeling in the presence of soft tissue and anatomical variability. By relaxing alignment assumptions and integrating vertebral pose estimation directly from image data, our method broadens the applicability of musculoskeletal simulations, particularly in cases with spinal deformities.

2 Method

We employed the point-of-interest (POI) generation code developed in Lerchl et al. [13] and evaluated the modifications required to adapt it for use with MRI. The method operates exclusively on segmentation and is independent of the imaging modality. Only the quality and resolution of the segmentation affect the algorithm. It depends on a detailed vertebra substructure segmentation, including the separation of anatomical subregions: vertebral body, arcus, spinous process, costal processes (left/right), superior articular processes (left/right), and inferior articular processes (left/right). For this purpose, we leveraged the open-access segmentation model SPINEPS [9,15], which is capable of producing such fine-grained segmentations from sagittal T2-weighted MRI. However, clinical sagittal T2w scans typically suffer from low in-plane resolution in the left-right direction (3–4 mm), due to practical constraints like acquisition time and the anatomical extent of the spine. When applying the original pixel-based methods by Lerchl et al. [13], we observed substantial inaccuracies under these conditions. Additionally, the method assumes that the vertebrae are aligned with the volume's left-right axis—a condition that may not hold in cases involving spinal deformities such as scoliosis. To improve robustness and anatomical accuracy, we introduce an algorithm that estimates the local orientation of each vertebra, decoupling the landmark computation from both voxel spacing and orientation. This approach replaces voxel-based assumptions with sub-pixel-accurate geometric logic, enabling consistent POI definition across varied scan orientations and resolutions. We release the full implementation under the Python package TPTBox, including tools to recompute POIs in alternative coordinate systems—such as different voxel spacings, ITK world space, or NIfTI world space—to support reproducibility and integration into broader workflows.

2.1 Vertebra Orientation

The original implementation used the cardinal directions of the image (left/right and front/back) as proxies for vertebral orientation. While this assumption holds for most healthy spines, it can break down in emergency settings, cases where patients cannot maintain a standard posture, or in the presence of spinal deformities such as scoliosis. We kept the original approach for computing the superior-inferior (up/down) direction: a spline is fitted through the centers of mass of the vertebral bodies, and the first derivative of this spline defines the local up/down direction. This approach avoids errors introduced by assuming that the vertebral body is cuboidal or that the endplates are flat and parallel.

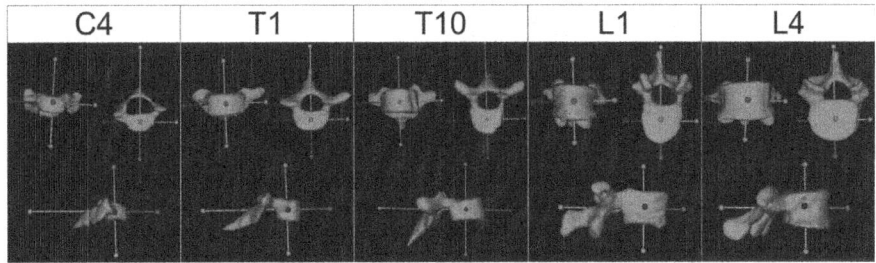

Fig. 3. Front, top, and right views of randomly selected vertebrae, visualized with the computed local coordinate system overlaid as directional whiskers. The top/bottom (cranio-caudal) axis is defined relative to adjacent vertebral bodies using a spline through their centers of mass. Due to anatomical asymmetries in structures such as the processus spinosus and arcus vertebrae, determining the front/back direction can be challenging. In particular, the L4 vertebra shown here exhibits significant asymmetry, which results in a slight rotation of the computed front/back direction compared to the visually expected anatomical posterior.

To compute the second anatomical direction, we extract the masks of the spinous process and the arcus vertebrae. These structures are projected onto a plane orthogonal to the up/down direction. The geometric centers of the projected masks are then computed, and a vector connecting the computed center with the center of mass of the vertebral corpus defines the second direction. The two vectors define a plane, and we recompute the front/back vector to be orthogonal to the up/down vector. The third direction is obtained as the cross product of the first (up/down) and second (front/back) directions, thereby forming an orthonormal local vertebral coordinate system (Fig. 3).

2.2 Ray Casts

The endpoints of all vertebral processes are now computed via raycasting from the center of mass of each corresponding subregion, using the previously computed local vertebral coordinate system (Fig. 4). The superior and inferior articular processes are defined along the superior-inferior axis (up and down directions, respectively). For the transverse processes, the raycasting direction is given by the vector

$$\mathbf{a} = 0.5 \cdot \mathbf{l} + 0.5 \cdot \mathbf{p},$$

where $\mathbf{l}$ is the lateral (left/right) direction and $\mathbf{p}$ is the posterior (backward) direction. For the spinous process, the raycasting direction is defined as

$$\mathbf{a} = \mathbf{d} + 0.2 \cdot \mathbf{p},$$

where $\mathbf{d}$ is the inferior (downward) direction and $\mathbf{p}$ is again the posterior direction. These direction vectors were empirically chosen to best match the observed

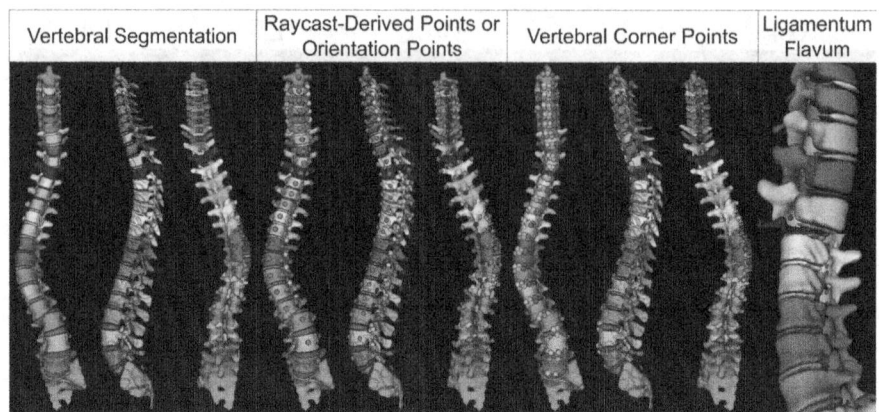

Fig. 4. Visualization of vertebral landmark extraction on a previously unseen sagittal T2-weighted MRI scan of a scoliotic spine. The first three panels show the vertebra segmentation overlaid with points, each with frontal, left, and back views. From left to right: segmentation only, the computed raycasting-based points, and corner landmarks. These highlight the robustness of the method under spinal curvature and low in-plane resolution. The final panel provides a close-up axial view of the spinal canal, focusing on the *ligamentum flavum* landmarks, which are accurately positioned on the posterior vertebral arch despite anatomical variability.

anatomical trajectories and to ensure robust point localization despite inter-subject variability.

The vertebral body is assigned one surface point in each of the six cardinal directions by raycasting from its center of mass along the corresponding direction vectors of the local vertebral coordinate system.

2.3 Vertebral Corpus Corners with Sub-voxel Bisection

In the original method, vertebral body corner points were extracted by selecting the sagittal slice intersecting the center of mass of the vertebral corpus. A Sobel filter was applied to detect edge candidates, and an image-aligned 2D bounding box was drawn around the vertebral body. The closest candidate points to the corners of this bounding box were then selected as corner landmarks. Additional intermediate edge points were computed and projected onto the vertebral surface.

In our updated approach, we computed the vertebral body midpoints previously via raycasting (Sect. 2.2). To compute the vertebral corpus corner points, we employ a 2D bisection search initialized at the center of mass and oriented along the local up/down and front/back direction vectors. The search iteratively steps outward in each direction and, upon exiting the segmentation boundary, the step size is halved repeatedly until a predefined precision threshold is reached. We interpolate at the tested coordinate to allow for subvoxel-accurate landmark placement that is robust to anatomical variability.

We compute the two additional anatomical landmarks corresponding to the *ligamentum flavum*, located in the same axial plane as the vertebral body corners but positioned on the anterior surface of the vertebral arch. The same 2D bisection strategy is used, starting from the center of mass of the arcus and constraining the search domain to the arcus segmentation mask rather than the corpus.

2.4 Shifted Points

The vertebral corners and cardinal direction landmarks are also computed using an offset in the left and right directions of the local vertebral coordinate system. These landmarks are anatomically motivated by the attachment of the anterior longitudinal ligament to the vertebral bodies. To improve anatomical accuracy, especially in the upper thoracic and cervical spine, we refined the original heuristic. Previously, the lateral shift was defined as one-third of the distance between the centers of mass of the superior articular processes. We now scale this shift using a vertebra-dependent factor

$$f = \frac{12 - v_{id}}{11} + 1 \quad \text{for} \quad v_{id} \leq 11,$$

where v_{id} is the vertebra index counted from the top, with C1 assigned as 1. This scaling accounts for the stronger shrinking of the vertebral bodies compared to the posterior structure in the neck and upper thoracic region.

3 Experiments and Results

3.1 Vertebra Orientation

To evaluate our new rotation estimation method, we randomly selected 90 vertebrae (from 20 subjects; 11 Female) from the VerSe2020 challenge dataset [18]. We manually measured the angular deviation between the estimated and true posterior directions. We report the mean angular deviation in degrees (°) and provide the fraction of results falling below two thresholds: one indicating excellent results and the other indicating catastrophic failures. A rotational deviation of $\leq 3°$ is rarely noticeable in qualitative inspection, so we use this as a practical threshold for an "excellent" orientation estimate. Conversely, deviations exceeding $10°$ were considered "catastrophic." The comparison in Table 1 highlights how often each method meets these criteria.

The original implementation did not include orientation estimation. We therefore experimented with several strategies that derive the vertebral coordinate system from the 3D centers of mass (CMS) of automatically extracted subregions. Owing to pronounced structural asymmetries and inter-subject anatomical variability, a naive 3D CMS of all posterior structures produced a front/back (anterior-posterior) direction that was off by $5.78 \pm 10.03°$ on average and exceeded $10°$ in 9% of the cases (8/90). This failure mode occurred almost exclusively in cervical levels, where the posterior elements are markedly skewed.

Table 1. Comparison of our proposed backward direction computation to a naive 3D center-of-mass (CMS) approach. Angles are measured in degrees (°). We report the mean ± standard deviation and the fraction of results with angular error below 3° and 10°.

	° Mean ± Std ↓	Fraction ≤ 3° ↑	Fraction ≤ 10° ↑
3D CMS (all posterior structures)	5.78 ± 10.03	0.39 (35/90)	0.91 (82/90)
3D CMS (Arcus and Spinosus)	2.87 ± 6.84	0.70 (63/90)	0.99 (89/90)
2D Projection (ours)	**1.72 ± 1.76**	**0.80 (72/90)**	**1.00 (90/90)**

Restricting the CMS to the arcus and spinosus parts mitigated this problem (2.87 ± 6.84°, with only 1/90 cases > 10°), yet the spread was still larger than we considered acceptable. To further stabilize the estimate, we introduced a regularization step: all relevant posterior voxels are first projected onto a 2D plane orthogonal to the superior-inferior axis, after which the 2D CMS is computed and re-embedded in 3D. This simple projection removes most of the out-of-plane asymmetry and shrinks the error to 1.72 ± 1.76°. Crucially, the method now achieves an error below 3° in 80% (72/90) and below 10° in 100% (90/90) of the vertebrae, which we deem sufficiently accurate for downstream shape analysis and visualization.

In summary, naively averaging all posterior voxels is prone to large angular errors, particularly in the cervical spine, whereas our 2D-projection strategy delivers robust, sub-3° accuracy in four out of five vertebrae and never exceeds 10°.

3.2 Points for Multi-body Simulation

To evaluate the reliability of our method for use in MBS, we tested it on 37 full spine segmentations. Two experts with 4 and 7 years of experience in Spine CT and MRI imaging assessed its performance. Failures occurred only in cases where the underlying segmentation was flawed.

We validated our point placement using an existing MBS framework [13]. Despite operating at a lower resolution, we observed no large discrepancy for straight spines, compared to existing CT-based point extraction. However, in some cases, we noticed large forces between the corner points of adjacent vertebral bodies. Upon investigation, we determined that this occurred when the vertebral bodies were closely aligned in the up/down direction but offset in the front/back or left/right direction. This issue arises from the definition of the frontal ligament used in the simulation, not from inaccuracies in the corner point placement. While this might indicate real tension in the anterior ligament, it is more likely due to the real anatomical attachment point located closer to the vertebral center. Accurately extracting the ligament path would be necessary to resolve this ambiguity, but this is not feasible with CT and is currently not available in MRI.

4 Discussion and Conclusion

We presented an advanced version of a rule-based POI extraction pipeline for the spine. The pipeline now supports MRI input, even with low left-right resolution, and can compensate for relative vertebral rotation—an essential capability for analyzing scoliotic spines. The used segmentation network and trained weights are publicly available, along with our enhanced point computation method. The entire POI computation completes in under a minute on a single CPU thread for a whole spine. Additionally, we provide tools for saving and loading the computed points and resampling them to different coordinate systems, such as voxel space, ITK, and NIfTI global space. The points can also be exported in the "mkr.json" format, allowing for easy import, editing, and visualization in 3D Slicer.

Our generated points provide a solid foundation for further development. While rule-based systems are effective, they tend to accumulate exceptions and special cases, such as fractured vertebrae, vertically misaligned segments, or the presence of metal implants, which become difficult to handle manually. In such scenarios, it is more efficient to correct the rule-based outputs and allow a deep learning model to generalize from them. Starting from our initial point annotations, it should be feasible to generate datasets that can be refined, corrected, and expanded, paving the way for robust, learning-based point prediction pipelines.

Acknowledgments. The research for this article received funding from the European Research Council (ERC) under the European Union's Horizon 2020 research and innovation program (101045128—iBack-epic—ERC2021-COG).

Our Code is available in the Python package https://github.com/Hendrik-code/TPTBox.

Disclosure of Interests. The authors have no competing interests to declare that are relevant to the content of this article.

References

1. Balasubramanian, S., D'Andrea, C.R., Viraraghavan, G., Cahill, P.J.: Development of a finite element model of the pediatric thoracic and lumbar spine, ribcage, and pelvis with orthotropic region-specific vertebral growth. J. Biomech. Eng. **144**(10), 101007 (2022)
2. Bürgin, V., Prevost, R., Stollenga, M.F.: Robust vertebra identification using simultaneous node and edge predicting graph neural networks. In: International Conference on Medical Image Computing and Computer-Assisted Intervention. pp. 483–493. Springer (2023)
3. Christophy, M., Curtin, M., Faruk Senan, N.A., Lotz, J.C., O'Reilly, O.M.: On the modeling of the intervertebral joint in multibody models for the spine. Multibody Syst. Dyn. **30**, 413–432 (2013)
4. Couvertier, M., et al.: Biomechanical analysis of the thoracolumbar spine under physiological loadings: experimental motion data corridors for validation of finite element models. Proc. Inst. Mech. Eng. [H] **231**(10), 975–981 (2017)

5. El Bojairami, I., El-Monajjed, K., Driscoll, M.: Development and validation of a timely and representative finite element human spine model for biomechanical simulations. Sci. Rep. **10**(1), 21519 (2020)
6. Fürmetz, J., et al.: Three-dimensional assessment of patellofemoral anatomy: reliability and reference ranges. Knee **29**, 271–279 (2021)
7. Fürmetz, J., et al.: Three-dimensional assessment of lower limb alignment: accuracy and reliability. KNEE **26**(1), 185–193 (2019)
8. Galbusera, F., Niemeyer, F., Bassani, T., Sconfienza, L.M., Wilke, H.J.: Estimating the three-dimensional vertebral orientation from a planar radiograph: is it feasible? J. Biomech. **102**, 109328 (2020)
9. Graf, R., et al.: Denoising diffusion-based MRI to CT image translation enables automated spinal segmentation. Eur. Radiol. Exp. **7**(1), 70 (2023)
10. Huynh, K., Gibson, I., Jagdish, B., Lu, W.: Development and validation of a discretised multi-body spine model in lifemod for biodynamic behaviour simulation. Comput. Methods Biomech. Biomed. Engin. **18**(2), 175–184 (2015)
11. Khanal, B., Dahal, L., Adhikari, P., Khanal, B.: Automatic cobb angle detection using vertebra detector and vertebra corners regression. In: International Workshop and Challenge on Computational Methods and Clinical Applications for Spine Imaging, pp. 81–87. Springer (2019)
12. Lachinov, D., Getmanskaya, A., Turlapov, V.: Cephalometric landmark regression with convolutional neural networks on 3d computed tomography data. Pattern Recognit Image Anal. **30**, 512–522 (2020)
13. Lerchl, T., et al.: Validation of a patient-specific musculoskeletal model for lumbar load estimation generated by an automated pipeline from whole body CT. Front. Bioeng. Biotechnol. **10**, 862804 (2022)
14. Lerchl, T., Nispel, K., Baum, T., Bodden, J., Senner, V., Kirschke, J.S.: Multi-body models of the thoracolumbar spine: a review on applications, limitations, and challenges. Bioengineering **10**(2), 202 (2023)
15. Möller, H., et al.: SPINEPS—automatic whole spine segmentation of t2-weighted MR images using a two-phase approach to multi-class semantic and instance segmentation. https://doi.org/10.1007/s00330-024-11155-y, https://doi.org/10.1007/s00330-024-11155-y
16. Nispel, K., Lerchl, T., Senner, V., Kirschke, J.S.: Recent advances in coupled MBS and fem models of the spine–a review. Bioengineering **10**(3), 315 (2023)
17. O'Neil, A.Q., et al.: Attaining human-level performance with atlas location autocontext for anatomical landmark detection in 3D CT data. In: Proceedings of the European Conference on Computer Vision (ECCV) Workshops. pp. 0–0 (2018)
18. Sekuboyina, A., et al.: Verse: a vertebrae labelling and segmentation benchmark (2020)
19. Tao, L., et al.: Automatic craniomaxillofacial landmarks detection in CT images of individuals with dentomaxillofacial deformities by a two-stage deep learning model. BMC Oral Health **23**(1), 876 (2023)
20. Wren, T.A., et al.: Biomechanical modeling of spine flexibility and its relationship to spinal range of motion and idiopathic scoliosis. Spine Deformity **5**(4), 225–230 (2017)
21. Zhang, H., Chung, A.C.: A dual coordinate system vertebra landmark detection network with sparse-to-dense vertebral line interpolation. Bioengineering **11**(1), 101 (2024)
22. Zhang, Y., et al.: Improving pedicle screw path planning by vertebral posture estimation. Phys. Med. Biol. **68**(18), 185011 (2023)

Towards Robust Algorithms for Surgical Phase Recognition via Digital Twin Representation

Hao Ding, Yuqian Zhang, Wenzheng Cheng, Xinyu Wang, Xu Lian, Chenhao Yu, Hongchao Shu, Ji Woong Kim, Axel Krieger, and Mathias Unberath[✉]

Johns Hopkins University, 3400 N Charles St, Baltimore, MD 21218, USA
{hding15,unberath}@jhu.edu

Abstract. Surgical phase recognition (SPR) is an integral component of surgical data science, enabling high-level surgical analysis. End-to-end trained neural networks that predict the surgical phase directly from videos have shown excellent performance on benchmarks. However, these models struggle with robustness due to non-causal associations in the training set. Our goal is to improve model robustness to variations in the surgical videos by leveraging the digital twin (DT) paradigm – an intermediary layer to separate high-level analysis (SPR) from low-level processing. As a proof of concept, we present a DT representation-based framework for SPR from videos. The framework employs vision foundation models with reliable low-level scene understanding to craft DT representation. We embed the DT representation in place of raw video inputs in the state-of-the-art SPR model. The framework is trained on the Cholec80 dataset and evaluated on out-of-distribution (OOD) and corrupted test samples. Contrary to the vulnerability of the baseline model, our framework demonstrates strong robustness on both OOD and corrupted samples, with a video-level accuracy of 80.3 on a highly corrupted Cholec80 test set, 67.9 on the challenging CRCD dataset, and 99.8 on an internal robotic surgery dataset, outperforming the baseline by 3.9, 16.8, and 90.9 respectively. We also find that using DT representation as an augmentation to the raw input can significantly improve model robustness. Our findings lend support to the thesis that DT representations are effective in enhancing model robustness. Future work will seek to improve the feature informativeness and incorporate interpretability for a more comprehensive framework.

Keywords: surgical video analysis · cholecystectomy · out-of-distribution generalization. domain generalization

1 Introduction

Surgical phase recognition (SPR) is a pivotal task in surgical data science, providing essential insights for surgical workflow analysis, skill assessment, and numerous other applications. With the rise of deep learning, end-to-end feed-forward networks [10,13,17,23,26,27] have demonstrated strong performance on SPR benchmarks [23,24]. Despite excellent performance, the end-to-end learning fashion for feed-forward networks (FFNs) poses the challenge of model robustness [9] when facing domain gap or non-adversarial corruptions [3]. The lack of reliability of the proposed algorithms hinders the clinical translation of research achievements for surgical data science [4,18]. Researchers started to take low-level vision tasks such as instrument segmentation and assess the models' robustness by adding corruptions to the input image [2]. Ding et al. added non-adversarial physical corruptions to the test data [3,6,7] ,taking advantage of the precise replayability of the da Vinci Research Kit [14]. However, to our best knowledge, there is a lack of exploration of the model's robustness for high-level analysis of surgical procedures like surgical phase recognition.

The digital twin (DT) paradigm, as illustrated in Fig. 1, constructs and maintains computational representations of real-world environments. It serves as an intermediary layer between the video pixels and SPR, achieving a separation between high-level analysis and low-level processing. From the real-world observations (physical twin) with high variance, DT applies semantic identification and geometric reconstruction techniques such as segmentation, depth estimation, 3D reconstruction, and pose estimation to extract a digital counterpart, a DT representation [4]. A high-level analysis model is then trained and applied on this low variance representation to reduce non-causal learning when training feed-forward networks, thereby enhancing the model's robustness. Nowadays, DT has received increasing attention, and extracting DT representations has become more feasible due to the emergence of vision foundation models for low-level processing, like the Segment Anything Model 2 (SAM2) [20] and DepthAnything [25]. These models have shown impressive zero-shot generalization and robustness [21], thanks to their large model capacity, vast training datasets, and advanced self- or semi-supervised learning mechanisms. Previous works [5,11,15,16,22] focus on the extraction of the DT representation from external tracking devices and markers or visual input. Ding et al. [5] applied the DT representation in an embodied surgical system and improved the robustness of the surgical automation, providing evidence of the effectiveness of the DT representation in improving the robustness of the downstream application.

In this work, we introduce a DT representationbased framework to improve robustness in SPR tasks. We harness the strong performance and robustness of vision foundation models by utilizing SAM2 [20] and DepthAnything [25] to extract basic DT representations. These structured representations then replace video inputs in the Surgformer [26] model, enabling SPR from surgical videos via the DT with enhanced robustness. We train our framework based on the Cholec80 [23] dataset and evaluate its performance and robustness on both original and corrupted images from the Cholec80 test dataset as well as two out-of-distribution (OOD) sets: CRCD [19] and an internal cholecystectomy dataset for robotics training acquired using the da Vinci Research Kit (dVRK) on ex vivo porcine specimens. Our results show that our framework demonstrates strong robustness against image corruptions and domain shifts, where the end-to-end

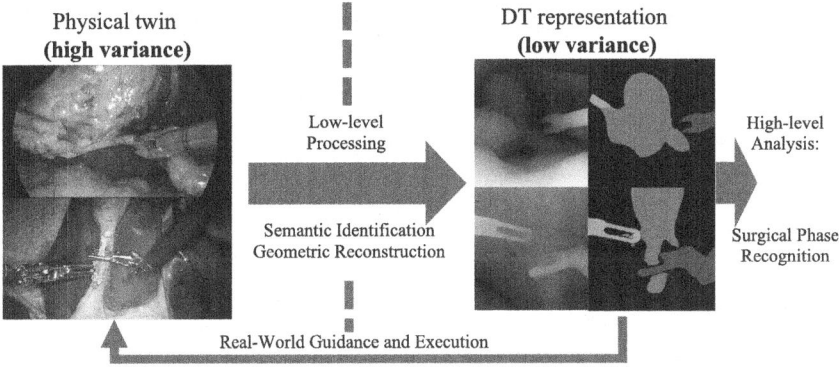

Fig. 1. Illustration of the DT paradigm. DT Paradigm demonstrates a clear separation between low-level processing with high-level analysis based on DT representation.

trained baseline models fail. We also successfully augmented the baseline model with the DT representation, achieving significantly improved robustness. These findings support the effectiveness of using DT representations to enhance model robustness, paving the way for more reliable high-level analysis for surgical data and accelerating the clinical translation of research outcomes. The experiment design and corresponding materials, like a 10-class instance segmentation of the full Cholec80 dataset, will also contribute to the following research.

Our contribution can be summarized as follows: (1) A robust SPR framework from surgical video via a DT representation; (2) An approach to form and apply a DT representation to state-of-the-art models; and (3) A comprehensive study of the robustness of the proposed framework on corrupted and OOD test data as well as the experimental materials for future research.

2 Methods

Figure 2 presents an overview of our proposed robust surgical phase recognition framework using DT representations. The framework comprises three primary modules: representation extraction, DT embedding, and DT-based SPR architecture - DT Former. In the representation extraction module, raw visual inputs, such as surgical videos, are processed using vision foundation models to extract explicit fundamental representations, including depth maps, segmentation masks, and their corresponding latent space tokens. These representations are further processed in the DT embedding module to form DT tokens. The final module, DT Former, applies the existing SPR model (Surgformer), taking DT tokens for training and inference. All modules will be detailed in the following subsections.

2.1 Representation Extraction

For representation extraction, we employ SAM2 [20] for instance segmentation and DepthAnything [25] for monocular depth estimation.

Instance Segmentation. We define 10 classes, as outlined in Table 1. These include one target tissue (gallbladder), one specimen bag, and six types of surgical instruments: grasper, bipolar, hook, scissors, clipper, and irrigator. To distinguish between multiple graspers that appear simultaneously, we differentiate them based on their insertion positions. For instance segmentation, we use SAM2 [20], providing a point prompt for each instance at the frame where it first appears. SAM2 utilizes a Vision Transformer (ViT) [8] as the image encoder and incorporates a memory attention module to integrate spatial and temporal information from previous and prompted frames. The point prompt is encoded using a prompt encoder, and a mask decoder processes the encoded prompt and image features to generate mask predictions while updating memory. In our framework, we take both the generated mask and the corresponding mask tokens output from SAM2's mask decoder and retain them for the following steps. The mask tokens for each frame are represented as 10 object entries, where each entry is a vector of length 257, representing one instance from Table 1. The first 256 dimensions of the vector are the output mask token from SAM2, where the last digit represents the existence of the object in the frame.

Depth Estimation. We apply the DepthAnything [25] model to each individual frame for depth estimation. The model follows an encoder-decoder architecture, estimating relative disparity from monocular images. The predicted disparity is normalized to a 0–1 scale, which we use to represent relative depth. The output tokens of the encoder and decoder are further processed by averaging spatial dimensions, producing the depth token used in subsequent steps.

Table 1. Instance list for segmentation. Each pair is the instance name and id.

gallbladder	1	left grasper	2	top grasper	3	right grasper	4	bipolar	5
hook	6	scissors	7	clipper	8	irrigator	9	specimen bag	10

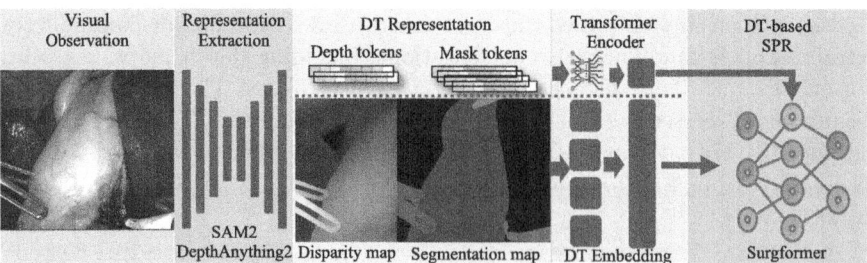

Fig. 2. Illustration of the surgical phase recognition framework via DT representation.

2.2 DT Embedding

For each frame, the raw representation consists of 10 binary segmentation masks for each class and a normalized disparity map. We format the segmentation masks into a 10-channel, one-hot encoded tensor, then concatenate the disparity map as the 11th channel. A 2D convolution layer is then applied to the sequence of frames, transforming the tensor into a spatial-temporal token sequence. To incorporate the depth token and the mask token, we applied two transformers for further processing. For the mask token, the first 256 dimensions of each 257-dimensional entry are projected into an embedding space, and a sinusoidal positional embedding is added to encode their positions. The embedded mask tokens are then appended by a learnable token and fed into self-attention blocks with the last digit of the entry as the mask for attention layers. The depth tokens undergo a similar embedding process, except without the binary mask for attention.

2.3 DT Former

We apply Surgformer [26] as the prediction model to determine the surgical phase for each frame. Surgformer processes a video clip of fixed length, appending a classification embedding to aggregate features for phase prediction. It integrates hierarchical temporal attention (HTA) into the TimesFormer [1] encoder to capture both global and local temporal information. The processed classification embedding from the final layer is passed to a linear head, predicting the phase of the last frame in online prediction settings. In our architecture, namely DT Former, we replace tokens from the patch embedding of the raw video clip with our DT tokens. The mask token is early fused with the CLS token via element-wise addition prior to the attention layers. By contrast, the depth token is late fused with the cls token via element-wise addition after the final layers.

3 Experiment

In this section, we present the experiments conducted to assess the robustness of our proposed method. We conduct two primary experiments: model robustness and ablation studies. For both experiments, our method and the baseline model are trained on the public Cholec80 dataset [23]. In the model robustness experiment, we evaluate our method on the original Cholec80 test set and corrupted Cholec80-C test set with combined corruption (Hue transformation, brightness, contrast), as well as the CRCD [19] and an internal robotics automation dataset, both serving as OOD test sets.

3.1 Experiment Setting

Dataset. We train our method and baselines on the Cholec80 dataset [23], which contains 80 cholecystectomy videos, each annotated with 7 surgical phases. The dataset is evenly split into training/validation (Video01-40) and testing sets (Video41-80). The CRCD [19] dataset includes videos recorded during ex vivo pseudo-cholecystectomy procedures on pig livers from which we take 5 videos performing gallbladder dissection for evaluation. The internal robotics automation dataset consists of 10 videos capturing a robotic ex vivo cholecystectomy procedure on the dVRK platform, specifically performing cutting and clipping tasks. All models are only trained on the training/validation split(Video01-40).

Evaluation Metrics. In this work, we evaluate performance using four key metrics: video-level accuracy, phase-level precision, recall, and Jaccard index. Video-level accuracy represents the mean percentage of correctly recognized frames for each video. Given the class imbalance, phase-level metrics offer further insights. Precision is defined as the ratio of true positives to total predictions, recall is the ratio of true positives to actual instances, and the Jaccard index measures the overlap between predictions and ground truths. For each phase, we collect the total number of true positives, false positives, and false negatives, and compute precision, recall, and Jaccard. Finally, we calculate the mean for these metrics across all phases. For the CRCD [19] and robotics automation test data, since they record specific procedures belonging to one specific phase for each dataset, we only report recall for phase-level metrics.

Model Setting. We apply the ViT-Tiny encoder for both the SAM2 [20] model and depthAnything [25] model and use the officially provided pre-trained model for inference. For the Surgformer [26] model, we adopt the official online SPR implementation with an input frame sequence length of 16, a sampling rate of 4, and a learning rate of 0.0005. We train each model for 30 epochs with a batch size of 8. The AutoAugmentation, random erasing, and random spatial sampling are also applied to the training of the baseline model.

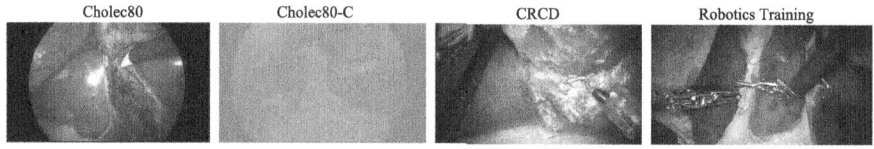

Fig. 3. Visual examples of the original and corrupted Cholec80 [23], CRCD [19], and the robotics automation dataset.

3.2 Model Robustness

OOD Generalization. As shown in Fig. 3, although these datasets depict the same procedure, there is a notable appearance gap between the Cholec80 [23], CRCD [19], and robotic cholecystectomy datasets, making them well-suited for OOD evaluation. The results, presented in the OOD section of Table 2, highlight that while the Surgformer baseline performs strongly on the Cholec80 test data, it struggles significantly on both OOD test sets, especially on the Robotics

Table 2. Robustness experiment results. Cholec80-C denotes the corrupted Cholec80. RA denotes robotics automation datasets. "D" and "M" means depth-only and mask-only.

OOD Methods	Cholec80 [23]				Cholec80-C [23]				CRCD [19]		RT	
	Acc	Prec	Rec	Jac	Acc	Prec	Rec	Jac	Acc	Rec	Acc	Rec
Surgformer	92.2	87.1	87.6	77.8	76.4	70.4	78.4	57.2	56.0	56.6	8.9	8.8
DT Former - D	83.4	75.3	77.7	62.3	37.3	50.3	40.9	20.6	38.4	36.1	99.1	99.1
DT Former - M	75.2	72.4	72.5	56.4	74.6	71.5	72.1	55.6	51.1	44.3	91.1	90.8
DT Former	85.5	79.9	78.2	67.2	**80.3**	**76.3**	**76.0**	**60.7**	**67.9**	**64.1**	**99.8**	**99.8**

Automation dataset, where it failed to identify cutting and clipping phases. By contrast, our model demonstrates robust performance, attaining 67.9 and 99.8 video-level accuracy, respectively on the CRCD [19] and robotics automation datasets, outperforming Surgformer by a large margin. These results underscore our model's robustness to OOD variations.

Robustness against Corruption. We apply combined corruption from hue transformation, brightness adjustment, and contrast alteration to the first 10 test videos to form a corrupted test set. The corruptions are selected and implemented according to the ImageNet-C practice [12]. Examples of these corrupted images are shown in Fig. 3. As seen in the corruption results in Table 2, the baseline model's performance substantially deteriorates under the corruption. In contrast, our framework demonstrates strong robustness overall, with only a slight drop in performance. We attribute this decline primarily to inaccuracies in the segmentation mask and depth prediction caused by the corruptions.

Table 3. Results of baseline model with DT augmentation.

Methods	Cholec80-C [23]				CRCD [19]		RA	
	Acc	Prec	Rec	Jac	Acc	Rec	Acc	Rec
Surgformer	76.4	70.4	78.4	57.2	56.0	56.6	8.9	8.8
Surgformer with DT-Aug	79.4	76.4	78.6	61.0	68.3	65.3	99.9	99.9

3.3 Ablation Study

For the ablation study, we first explore the contribution of depth and segmentation information in the DT Former to assess the contribution of depth and segmentation. Then we explore the possibility of using DT representation as an augmentation to raw video input.

Components. We create depth-only and segmentation-only models by removing one modality at a time from the input. The results, presented in Table 2, show that, between the two variants, taking depth information solely as input generates better benchmark performance, but solely taking mask information as input generates better robustness. We suppose this is because depth representation contains more fine-grained details, making the learning process easier, while the mask information is more abstract; thus, the performance of the corresponding foundation models is more consistent when facing corruptions. Combining a mask and depth makes a good complement for each other, improving the comprehensive model performance.

DT Representation Augmented Baseline. DT representation-based model demonstrates strong robustness in the previous experiment. In this study, we explore another possibility of incorporating the DT representation as an augmentation into the raw input to improve the baseline model's robustness. We concatenate our 11-channel spatial DT input with the raw RGB input and adapt the depth and history tokens into the baseline model to form the augmented baseline. The results, presented in Table 3 show that the DT augmentation improves the robustness of the baseline model on both corrupted and OOD test samples. This further validates the idea of applying DT representation for better model robustness in high-level analysis tasks.

4 Discussion

Our DT-based SPR framework demonstrates strong robustness in high-level surgical analysis tasks, underscoring the efficacy of the DT paradigm as an intermediary layer for enhancing surgical data science research. The advent of vision foundation models provides a highly generalizable and reliable low-level processing pipeline, facilitating the extraction and formation of DT representations. This capability holds promise for fostering more generalizable and interpretable systems in surgical data science, potentially accelerating the clinical translation of these research advancements. As proof of concept and an initial exploration of DT-based high-level analysis, further development is required to create a more comprehensive and practical framework for SPR and other high-level analytical tasks. On one hand, despite using the same model capacity, on clean data, the benchmark performance of our framework remains below the state-of-the-art methods, indicating that it may not yet extract sufficient information for SPR. Thus, Future research should investigate additional low-level features, such as scene flow or action triplet, which can provide informative data for SPR and be reliably extracted. We believe there isn't necessarily a trade-off between performance on clean data and robustness, as eventually, with adequate structured modality in the digital twin representation, we can achieve comparable performance on the cleaned benchmark test set while also improving the performance on OOD and corrupted test samples. On the other hand, we believe inferencing on the interpretable digital twin representation can also improve the interpretability of the inference of the final results. For example, we can do counterfactual inference by altering digital twin representations (e.g., changing the tool label) to measure the direct causal effect in the model inference. This interpretability offered by the DT representation has not yet been fully explored. Such interpretability could be highly relevant to clinical settings by clarifying decision-making processes and instilling greater trust in automated systems, and therefore is a direction to further explore in future efforts. Another unignorable limitation ithat s the use of foundation models has brought the concern on inference speed. We believe this will eventually not be an issue with the development of the power of the hardware and the efficiency of the foundation models.

5 Conclusion

We propose a surgical video analysis framework using a DT representation. This framework demonstrates strong robustness against OOD and corrupted test samples, validating the effectiveness of high-level surgical analysis via DT representation and the reliability of low-level representation extraction through vision foundation models. Our work highlights a promising research direction for developing more robust algorithms in surgical data science, moving beyond recent end-to-end deep learning approaches. To create a more comprehensive framework, future research should focus on enhancing the informativeness of DT-based feature representations and improving the efficacy of the feature extraction pipeline. Additionally, incorporating explainable AI techniques to enhance interpretability is likely beneficial for facilitating clinical translation.

Acknowledgments. his research is supported by a collaborative research agreement with the MultiScale Medical Robotics Center at The Chinese University of Hong Kong.

Disclosure of Interests. The authors have no competing interests to declare that are relevant to the content of this article.

References

1. Bertasius, G., Wang, H., Torresani, L.: Is space-time attention all you need for video understanding?. In: ICML, vol. 2, p. 4 (2021)
2. Colleoni, E., Edwards, P., Stoyanov, D.: Synthetic and real inputs for tool segmentation in robotic surgery. In: International Conference on Medical Image Computing and Computer-Assisted Intervention, pp. 700–710. Springer (2020)
3. Ding, H., et al.: SegSTRONG-C: segmenting surgical tools robustly on non-adversarial generated corruptions–an Endovis' 24 challenge. arXiv preprint arXiv:2407.11906 (2024)
4. Ding, H., Seenivasan, L., Killeen, B.D., Cho, S.M., Unberath, M.: Digital twins as a unifying framework for surgical data science: the enabling role of geometric scene understanding. Artif. Intell. Surge. 4(3), 109–138 (2024)
5. Ding, H., et al.: Towards robust automation of surgical systems via digital twin-based scene representations from foundation models. arXiv preprint arXiv:2409.13107 (2024)

6. Ding, H., Wu, J.Y., Li, Z., Unberath, M.: Rethinking causality-driven robot tool segmentation with temporal constraints. Int. J. Comput. Assist. Radiol. Surg. **18**(6), 1009–1016 (2023)
7. Ding, H., Zhang, J., Kazanzides, P., Wu, J.Y., Unberath, M.: Carts: causality-driven robot tool segmentation from vision and kinematics data. In: Proc. MICCAI, pp. 387–398. Springer (2022)
8. Dosovitskiy, A., et al.: An image is worth 16×16 words: transformers for image recognition at scale. In: Proc. ICLR (2021)
9. Drenkow, N., Sani, N., Shpitser, I., Unberath, M.: A systematic review of robustness in deep learning for computer vision: mind the gap?. arXiv preprint arXiv:2112.00639 (2021)
10. Gao, X., Jin, Y., Long, Y., Dou, Q., Heng, P.: Trans-SVNet: accurate phase recognition from surgical videos via hybrid embedding aggregation transformer. In: Proc. MICCAI (2021)
11. Hein, J., et al.: Creating a digital twin of spinal surgery: a proof of concept. In: Proc. CVPR, pp. 2355–2364 (2024)
12. Hendrycks, D., Dietterich, T.: Benchmarking neural network robustness to common corruptions and perturbations. arXiv preprint arXiv:1903.12261 (2019)
13. Jin, Y., Dou, Q., Chen, H., Yu, L., Qin, J., Fu, C., Heng, P.: SV-RCNet: workflow recognition from surgical videos using recurrent convolutional network. IEEE Trans, Med. Imag. **37**(5) (2018)
14. Kazanzides, P., Chen, Z., Deguet, A., Fischer, G.S., Taylor, R.H., DiMaio, S.P.: An open-source research kit for the da Vinci® Surgical System. In: Proc. ICRA, pp. 6434–6439. IEEE (2014)
15. Killeen, B.D., et al.: Stand in surgeon's shoes: virtual reality cross-training to enhance teamwork in surgery. In: Int. J. CARS, pp. 1–10 (2024)
16. Kleinbeck, C., Zhang, H., Killeen, B.D., Roth, D., Unberath, M.: Neural digital twins: reconstructing complex medical environments for spatial planning in virtual reality. Int. J. CARS **19**, 1–12 (2024)
17. Liu, Y., et al.: SKiT: a fast key information video transformer for online surgical phase recognition. In: Proc. ICCV (2023)
18. Maier-Hein, L., et al.: Surgical data science-from concepts toward clinical translation. Med. Image Anal. **76**, 102306 (2022)
19. Oh, K.H., et al.: Comprehensive robotic cholecystectomy dataset (CRCD): integrating kinematics, pedal signals, and endoscopic videos. In: Proc ISMR (2024)
20. Ravi, N., et al.: SAM 2: segment anything in images and videos. arXiv preprint arXiv:2408.00714 (2024)
21. Shen, Y., Ding, H., Shao, X., Unberath, M.: Performance and non-adversarial robustness of the segment anything model 2 in surgical video segmentation. arXiv preprint arXiv:2408.04098 (2024)
22. Shu, H., et al.: Twin-s: a digital twin for skull base surgery. Int. J. CARS **18**(6), 1077–1084 (2023)
23. Twinanda, A.P., Shehata, S., Mutter, D., Marescaux, J., de Mathelin, M., Padoy, N.: EndoNet: a deep architecture for recognition tasks on laparoscopic videos. IEEE Trans, Med. Imag. **36**(1) (2017)
24. Wang, Z., et al.: AutoLaparo: a new dataset of integrated multi-tasks for image-guided surgical automation in laparoscopic hysterectomy. In: Proc. MICCAI (2022)
25. Yang, L., Kang, B., Huang, Z., Xu, X., Feng, J., Zhao, H.: Depth anything: unleashing the power of large-scale unlabeled data. In: Proc. CVPR, pp. 10371–10381 (2024)

26. Yang, S., Luo, L., Wang, Q., Chen, H.: SurgFormer: surgical transformer with hierarchical temporal attention for surgical phase recognition. In: Proc. MICCAI, pp. 606–616. Springer (2024)
27. Yi, F., Jiang, T.: Hard frame detection and online mapping for surgical phase recognition. In: Proc. MICCAI (2019)

Personalized 4D Whole Heart Geometry Reconstruction from Cine MRI for Cardiac Digital Twins

Xiaoyue Liu[1], Xicheng Sheng[2], Xiahai Zhuang[2], Vicente Grau[3], Mark YY Chan[4,5], Ching-Hui Sia[4,5], and Lei Li[1(✉)]

[1] Department of Biomedical Engineering, National University of Singapore, Singapore, Singapore
lei.li@nus.edu.sg
[2] School of Data Science, Fudan University, Shanghai, China
[3] Department of Engineering Science, University of Oxford, Oxford, UK
[4] Department of Medicine, National University of Singapore, Singapore, Singapore
[5] Department of Cardiology, National University Heart Centre Singapore, Singapore, Singapore

Abstract. Cardiac digital twins (CDTs) provide personalized in-silico cardiac representations and hold great potential for precision medicine in cardiology. However, whole-heart CDT models that simulate the full organ-scale electromechanics of all four heart chambers remain limited. In this work, we propose a weakly supervised learning model to reconstruct 4D (3D+t) heart mesh directly from multi-view 2D cardiac cine MRIs. This is achieved by learning a self-supervised mapping between cine MRIs and 4D cardiac meshes, enabling the generation of personalized heart models that closely correspond to input cine MRIs. The resulting 4D heart meshes can facilitate the automatic extraction of key cardiac variables, including ejection fraction and dynamic chamber volume changes with high temporal resolution. It demonstrates the feasibility of inferring personalized 4D heart models from cardiac MRIs, paving the way for an efficient CDT platform for precision medicine. The code will be publicly released once the manuscript is accepted.

Keywords: Cine MRI · 4D Heart Reconstruction · Whole Heart · Self-Supervised Mapping · Cardiac Digital Twins

1 Introduction

Cardiovascular diseases remain the leading cause of mortality worldwide. Cardiac digital twin (CDT) technology has emerged as a powerful approach for creating patient-specific virtual heart models, enabling real-time analysis of cardiac structure and function [15]. By offering detailed insights into the underlying mechanisms of the heart, CDT has the potential to revolutionize cardiac diagnosis and treatment [1,9]. A typical CDT workflow consists of two key stages: anatomical twinning and functional twinning [10,14]. Anatomical twinning involves extracting 3D heart geometry from images and identifying pathological regions when present. Considering the dynamic nature of cardiac motion, 4D (3D+t) geometry is typically required for a more comprehensive representation. Cine MRI can be used for this purpose, as it provides non-invasive visualization of cardiac anatomy and motion throughout the cardiac cycle. However, cine MRI typically acquires sparse and intersecting 2D image planes, i.e., short-axis (SAX) and long-axis (LAX) slices, which limits spatial resolution and hinders the construction of a fully detailed 4D representation of the heart. These constraints can impact the accuracy of anatomical twinning and downstream functional twinning which involves the simulation of cardiac electromechanics.

Conventional cardiac geometry reconstruction frameworks generally consist of two steps, i.e., image segmentation, followed by mesh generation based on contours derived from the segmentation. This is mainly because direct volumetric reconstruction from cine MRIs is challenging due to the inherent sparsity and anisotropy of the data. By first segmenting the cardiac structures, the extracted contours can serve as geometric constraints to guide the mesh generation. However, cine MRI only provides a sparse representation of the actual 3D geometry of the human heart. Consequently, traditional iso-surfacing algorithms, such as marching cubes, struggle to generate smooth and anatomically accurate meshes due to the irregular spacing and insufficient volumetric information in the input data. To solve this, many previous studies employed mesh adaptation approaches, such as template mesh deformation [11,18], statistical shape model (SSM) [2], and B-spline surface reconstruction [4,16]. Furthermore, image interpolation based methods have also been applied to reconstruct high-resolution 3D geometry [17]. Nonetheless, these techniques are labor-intensive, requiring complex manual initialization or adjustments of optimizer parameters, which significantly hinders their feasibility for real-time applications.

Recently, deep learning-based methods have achieved promising performance for efficient 3D cardiac geometry reconstruction. Similarly, they generally rely on pre-generated segmentation and then directly convert sparse contour point clouds into 3D meshes via point completion network [3] or graph convolution network (GCN) based template deformation [7,19]. Instead of directly performing interpolation on images, Chang et al. [6] developed a latent space-based generative method to simultaneously predict 2D SAX segmentation and 3D volume by interpolating latent codes. Biffi et al. [5] first segmented cine MRI and then reconstructed a high-resolution 3D volume using a conditional variational autoencoder, incorporating features from one SAX and two LAX segmentation. For 4D cardiac reconstruction, Yuan et al. [20] decoupled cardiac motion and shape from the given sparse contour point cloud sequences based on the neural motion model and deep SSM model. Recently, there are several deep learning based studies directly reconstruct meshes from cardiac image [8,12,13]. For example, Kong et al. [12] directly predicted whole heart surface meshes from volumetric CT and MRI data via GCN based pre-defined mesh template deformation. Chen et al. [8] employed a deep marching tetrahedra model that discretized 3D space into a deformable tetrahedral grid, assigning each vertex a signed distance value for 4D biventricular reconstruction. Laumer et al. [13] reconstructed 4D whole heart mesh from 2D echocardiography video data via task-tailored autoencoder models. However, their approach relied on single-view, single-slice 2D data, which provided limited spatial information of the heart and thus constrained the reconstruction accuracy. In general, current work either relied on high-resolution volumetric images or solely reconstructed part of the whole heart.

In this work, we present a novel weakly supervised model to infer personalized whole-heart meshes from multi-view cine MRI. Given unpaired cine MRI and 4D heart mesh, the model can efficiently map the cine MRI to a 4D heart mesh via the generative domain translation. Specifically, we utilize the domain-specific autoencoder networks to extract the compact latent representations of both cine MRIs and the mesh videos. Then, the mapping between the image and heart mesh video latent spaces can be learned to ensure the generated shapes align with the cardiac deformation space. To the best of our knowledge, this is the first study to directly reconstruct a 4D whole-heart mesh from cine MRIs.

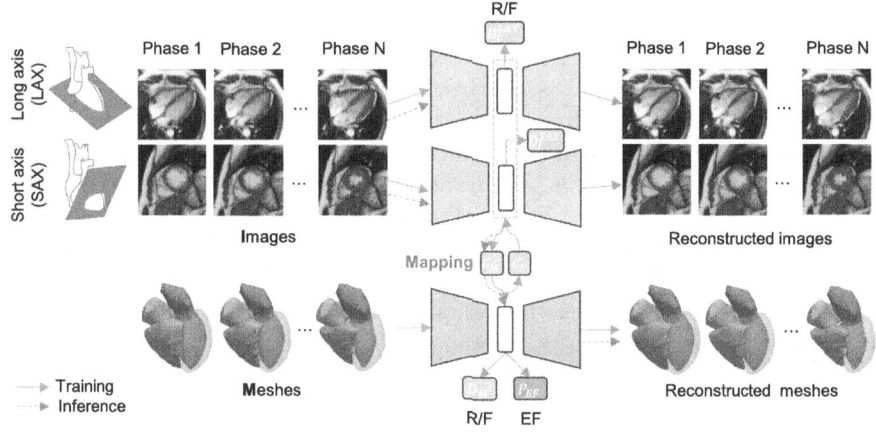

Fig. 1. Illustration of the multi-view cine-MRI based 4D whole heart mesh reconstruction framework. Note that the diagram only shows a single short-axis (SAX) slice as an example, even though three representative slices in SAX view are selected to capture comprehensive spatial information.

2 Methodology

Figure 1 provides an overview of the proposed 4D whole heart reconstruction model, consisting of domain-specific autoencoders (AE) and cycle-mapping modules. Image and mesh AE networks are designed to extract the compact latent representations of cine MRIs and mesh videos, respectively (Sect. 2.1). To train the generator network mapping between the image and heart mesh video latent space, we optimize a cycle-consistency loss to enforce reversibility and an adversarial loss with two discriminators distinguishing real from fake representations (Sect. 2.2). Additionally, ejection fraction predictor acts as a weak regularizer to guide generator training. Finally, Sect. 2.3 presents the details of the reconstruction model for the personalized inference of 4D whole heart mesh.

2.1 Domain-Specific Autoencoder for Feature Extraction

To extract the cardiac spatial and motion features from cine MRIs, we employ the image-specific AE (namely AE_I) on two different cine views, i.e., LAX and SAX views. LAX view includes single slice, while SAX view is the stack of several slices but we only select three representative slices, i.e., apical, middle, and basal slices, for simplification. Given a sequence of cine frames, $\{I_t^{SAX}, I_t^{LAX}\}_{t=1}^{N}$, where t is the frame ID and N is the number of frames in the cine data, we can extract the corresponding low-dimensional manifold $\mathcal{Z}_I$ for each view and fuse them as cine imaging latent space. To achieve this, we integrate prior knowledge of periodicity by representing the latent trajectory as a circular motion in the first two dimensions, with additional parameters controlling shape and dynamics. The AE_I encoder, consisting of a CNN and an LSTM, extracts features and predicts trajectory parameters ϕ_I as its final state. Instead of directly using these parameters, AE_I decoder reconstructs observations based on the learned latent trajectory to ensure temporal consistency. The optimization of AE_I is based on minimizing a regularized image reconstruction loss:

$$\mathcal{L}_I^{recon} = \frac{1}{N} \sum_{t=1}^{N} \|I_t - \hat{I}_t\|^2 + \mathcal{R}(\phi_I), \tag{1}$$

where $I_j t$ and $\hat{I}_t$ are the input and reconstructed cine frames, respectively, and $\mathcal{R}(\phi_I)$ is the regularizer to constrain the latent trajectory parameters.

Given a mesh video as a sequence of 3D heart meshes with corresponding time steps, represented as $\{M_t\}_{t=1}^{N}$, we can similarly extract their spatial and motion features via a mesh-specific AE (namely AE_M). It consists of an encoder that encompasses a feature extractor network and an LSTM which outputs as its final state compressed representation ϕ_M. The feature extractor and the AE_M decoder is based on graph neural networks. This is because each surface mesh M_t consists of vertices and faces forming polygonal surfaces and can be described as an undirected graph $G = (\mathcal{V}, \mathcal{E})$, where $\mathcal{V} = \{v_1, ..., v_n\}$ is the set of n vertices, and $\mathcal{E} = \{(v_i, v_j)\}$ (for $i \neq j$) represents the set of edges connecting the vertices. The neighborhood of a vertex v_i, denoted as $\mathcal{N}_i$, is defined as the set of vertices v_j such that $(v_i, v_j) \in \mathcal{E}$, i.e., $\mathcal{N}_i = \{v_j \mid (v_i, v_j) \in \mathcal{E}\}$. The graph structure is represented by an adjacency matrix A and vertex v_i is associated with a feature vector. The optimization of AE_M is achieved by minimizing a regularized mesh reconstruction loss:

$$\mathcal{L}_M^{recon} = \frac{1}{N} \sum_{t=1}^{N} \|M_t - \hat{M}_t\|^2 + \mathcal{R}(\phi_M), \tag{2}$$

where M_jt and $\hat{M}_t$ are input and reconstructed mesh video, respectively, and $\mathcal{R}(\phi_M)$ is the regularization term.

2.2 Domain Translation Based Cycle Feature Mapping

We assume that latent representations of cine imaging and mesh videos lie on separate manifolds, $\mathcal{Z}_I$ and $\mathcal{Z}_M$, with mapping functions $G_M : \mathcal{Z}_I \to \mathcal{Z}_M$ and $G_I : \mathcal{Z}_M \to \mathcal{Z}_I$. To estimate the corresponding 4D whole heart mesh for cine MRI, we need to perform a domain translation. To achieve this, we employ generative adversarial networks to find the mapping between the two domains. Specifically, the generator G_M learns to transform echo representations into realistic mesh video representations, while the discriminator D_M distinguishes real from generated samples, optimized via adversarial loss:

$$\mathcal{L}_M^{\text{adv}} = \mathbb{E}_{\phi_I}[\log D_M(G_M(\phi_I))] + \mathbb{E}_{\phi_M}[\log(1 - D_M(\phi_M))], \tag{3}$$

$$\mathcal{L}_I^{\text{adv}} = \mathbb{E}_{\phi_M}[\log D_I(G_I(\phi_M))] + \mathbb{E}_{\phi_I}[\log(1 - D_I(\phi_I))], \tag{4}$$

where ϕ_I and ϕ_M are latent representations of cine and mesh videos, respectively. To ensure consistency between the mapping functions G_I and G_M, we employ a cycle-consistency loss:

$$\mathcal{L}^{\text{cycle}} = \mathbb{E}_{\phi_I}\left[\|G_I(G_M(\phi_I)) - \phi_I\|_1\right] + \mathbb{E}_{\phi_M}\left[\|G_M(G_I(\phi_M)) - \phi_M\|_1\right], \tag{5}$$

where $\|\cdot\|_1$ denotes the L1 norm.

To further improve the correspondences between cine and mesh videos, we introduce a pretrained EF prediction network N_{EF} to capture meaningful shape and dynamics information. The ejection fraction (EF) is determined using the end-diastolic (ED) volume and end-systolic (ES) volume. During training, we predict the EF from generated mesh representations $G_M(\phi_I)$ and $G_M(G_I(\phi_M))$ using N_{EF}. The EF loss is defined as:

$$\mathcal{L}^{\text{EF}} = \mathbb{E}_{\phi_I}\left[\|N_{\text{EF}}(G_M(\phi_I)) - \text{EF}_I\|_1\right] + \mathbb{E}_{\phi_M}\left[\|N_{\text{EF}}(G_M(G_E(\phi_M))) - \text{EF}_M\|_1\right], \tag{6}$$

where EF_I and EF_M are the ground-truth EF values from the cine and mesh data, respectively.

2.3 Personalized Inference of 4D Whole Heart Mesh

Two domain-specific AEs can be independently trained, after which their weights are fixed. Subsequently, we can extract the compressed representations ϕ_I and ϕ_M from cine and mesh video data to train the domain shift modules. At the same time, the EF prediction network N_{EF} is trained on $\mathcal{D}_M$. Therefore, the mapping networks G_M, G_E, and discriminators D_E, D_M are optimized based on the following objective:

$$\min_{\lambda_{D_M}, \lambda_{D_E}} \max_{\lambda_{G_M}, \lambda_{G_E}} \beta_1 \mathcal{L}_M^{\text{adv}} + \beta_2 \mathcal{L}_E^{\text{adv}} + \beta_3 \mathcal{L}^{\text{cycle}} + \beta_4 \mathcal{L}^{\text{EF}}, \tag{7}$$

where λ_{D_M}, λ_{D_E}, λ_{G_M}, and λ_{G_E} are the balancing parameters of the respective networks. Expectations in the loss functions are estimated via mini-batch sampling. Training stops when $\mathcal{L}_{\text{EF}}$ on the validation sets stops decreasing. During inference, the weights of all networks are fixed, and mesh videos are generated for each cine data using:

$$\hat{\mathcal{M}} = \text{AE}_M^{\text{Decoder}}(G_M(\text{AE}_I^{\text{Encoder}}(I^{LAX}, I^{SAX}))). \tag{8}$$

This process aims to generate personalized 4D meshes for each cine data.

3 Experiments and Results

3.1 Materials

Data Acquisition and Pre-processing. We collected 446 subjects with standard multi-view cardiac cine MRI. Specifically, three LAX slices and a SAX stack of 610 slices over one cardiac cycle are existed. In total, 25–50 frames per cardiac cycle were obtained for each subject in the study population. We only utilize the 4-chamber LAX and three representative slices from the SAX cine data. All images were cropped into a unified size of 112 × 112 centering at the heart region, with a intensity normalization via Z-score. The dataset was randomly divided into 313 training and 133 test samples. To train the mesh-specific AE, we generated 10,000 whole-heart surface mesh samples based on the SSM which embeds the morphological variation and dynamics observed in a cohort of 20 patients.

Implementation. The framework was implemented in TensorFlow, running on a computer with a 13th Gen Intel(R) Core(TM) i9-13980HX CPU and an NVIDIA GeForce RTX 4060 Laptop GPU. We used the Adam optimizer to update the network parameters via stochastic gradient decent. The balancing parameters in Sect. 2.3 are set as follows: $\beta_1 = 1$, $\beta_2 = 1$, $\beta_3 = 10$, and $\beta_4 = 10$. The training of the model took about 65 h, while the inference of one 4D heart from input cine images required about 18 mins.

Gold Standard and Evaluation. Cine MRIs were manually segmented by a well-trained student using ITK-SNAP and checked by a senior expert. For the SAX view, LV, LV Myo, and RV have been labeled, while for the LAX view, LV, LV Myo, RV, LA, RA have been labeled. These manual labels have been converted into contours, which are considered as ground truth in this work. For evaluation, we employed average surface distance (ASD) to assess the alignment between the predicted heart and the corresponding contours. Note that since the coordinate information was not learned in the generated mesh, we performed an initial alignment based on iterative closest point (ICP) registration algorithm between the contour and generated mesh for validation.

Table 1. Summary of the quantitative evaluation results of 4D heart reconstruction in terms of average surface distance (mm).

View	Phase	Myo	LV	RV	LA	RA	Avg
LAX	ED	6.12 ± 1.39	5.34 ± 1.14	10.7 ± 2.29	7.19 ± 1.32	6.08 ± 1.39	7.08
	ES	7.66 ± 1.83	6.55 ± 1.46	9.28 ± 0.90	5.97 ± 0.98	5.38 ± 1.07	6.97
LAX+SAX	ED	5.39 ± 0.94	5.66 ± 1.98	11.3 ± 2.74	7.20 ± 1.60	5.69 ± 1.24	7.06
	ES	6.41 ± 1.52	5.60 ± 0.91	9.72 ± 1.55	5.82 ± 1.04	5.80 ± 1.21	6.67

3.2 Results

Effect of the Number of Cine Views. To investigate the impact of the number of input views, we compared the performance of the proposed model using a single LAX view with that of a model incorporating both SAX and LAX views. We randomly selected 20 subjects and manual segmented their LAX and SAX view at the ED and ES phases. Table 1 presents the average ASD values between reconstructed mesh and ground truth contours of the 20 subjects. One can see that in general combining multi-view cine MRI can perform slightly better than using single LAX view (ED: 7.06 mm vs. 7.08 mm and ES: 6.67 mm vs. 6.97 mm). This is reasonable, as incorporating both views provided complementary spatial information for capturing complex cardiac structures and motion patterns across different perspectives. Multi-view fusion also mitigated errors or ambiguities from a single view, leading to more robust reconstructions. However, the improvement remained limited, as the model relies on learned priors and shape constraints, reducing sensitivity to additional views. Also, the weakly supervised nature of the model prevented explicit learning of anatomical correspondences, limiting the benefits of multi-view integration.

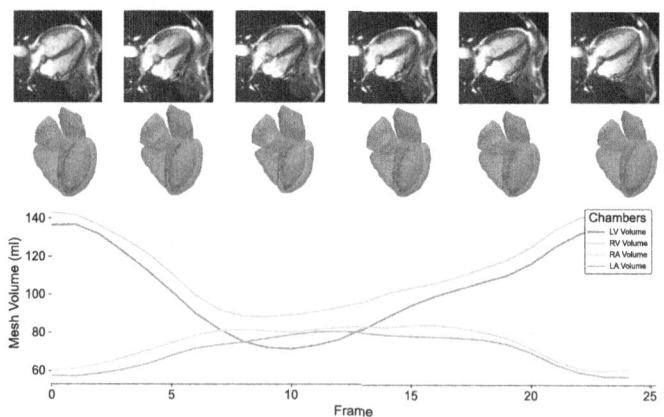

Fig. 2. Illustration of cine MRI and predicted whole heart mesh with corresponding four chamber volume change over time.

Accuracy of the Reconstructed 4D Heart. Figure 2 presents an example of the predicted heart shapes alongside the corresponding chamber volume changes over time. One can see that the reconstructed meshes accurately captured the cardiac systolic and diastolic phases observed in the cine MRI, demonstrating effective preservation of cardiac motion dynamics. Additionally, Fig. 3 (a) shows the overlap visualization of predicted mesh and the ground truth in three randomly selected subjects. One can see that the predicted chamber volume changed generally align with the ground truth, further validating the capability of the proposed model to adapt to both the shape and motion patterns of the heart. This consistency suggests that the learned representations effectively encode physiological deformations, enabling accurate and realistic 4D reconstructions.

Capability of the Reconstructed 4D Heart for EF Estimation. To further assess the accuracy of the reconstructed mesh, we compared EF values derived from manual segmentation of cine MRIs and predicted mesh, visualizing as a scatter plot for all test subjects (see Fig. 3 (b)). To quantify this correlation, we performed linear regression and Pearson correlation analyses. The Pearson correlation coefficient of 0.563 indicated a moderate to strong correlation, suggesting that the reconstructed 4D heart model effectively captured volumetric changes during the cardiac cycle. However, some discrepancies may arise due to limited diversity of cardiac morphology and motion dynamics in the reconstruction process, which could restrict its ability to fully capture patient-specific variations. Further refinement, such as incorporating additional anatomical constraints, may enhance the accuracy of EF estimation.

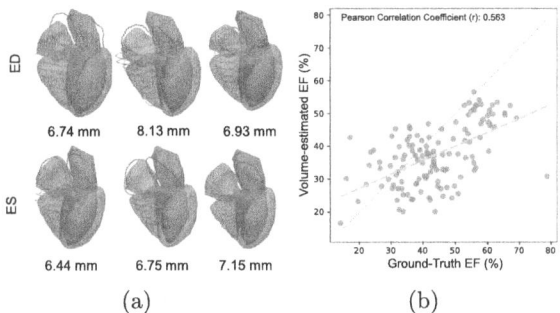

Fig. 3. (a) 3D visualization of the overlap between the sparse contours and reconstructed heart; (b) The scatter point and correlations between the ground truth and estimated EF.

4 Conclusion

In this work, we have proposed an end-to-end learning framework for automatic 4D whole heart mesh reconstruction by combining the LAX and SAX cine

MRIs. The proposed algorithm has been applied to 446 subjects and obtained promising results compared manual delineation. The results have demonstrated the effectiveness of the proposed mapping scheme and showed the feasibility of employing 2D(+t) images for 3D(+t) reconstruction. Particularly, the proposed model does not rely on any paired imaging and mesh data nor the corresponding manual mask of cine MRIs. Despite its effectiveness, this work has three main limitations. First, the input mesh distribution is derived from a small cohort of 20 subjects, limiting its ability to capture diverse cardiac shapes and motion patterns. Second, the reconstructed mesh lacks original image coordinates, which could be addressed using spatial constraints like cardiac landmarks or segmentation priors. Finally, further investigation is needed to validate the integration of the 4D mesh into cardiac simulations for efficient digital twin construction.

Acknowledgement. This work was supported by NUS start-up funding to L. Li.

References

1. Arevalo, H.J., et al.: Arrhythmia risk stratification of patients after myocardial infarction using personalized heart models. Nat. Commun. **7**(1), 11437 (2016)
2. Banerjee, A., et al.: A completely automated pipeline for 3D reconstruction of human heart from 2D cine magnetic resonance slices. Phil. Trans. R. Soc. A **379**(2212), 20200257 (2021)
3. Beetz, M., Banerjee, A., Ossenberg-Engels, J., Grau, V.: Multi-class point cloud completion networks for 3D cardiac anatomy reconstruction from cine magnetic resonance images. Med. Image Anal. **90**, 102975 (2023)
4. Bennati, L., et al.: Turbulent blood dynamics in the left heart in the presence of mitral regurgitation: a computational study based on multi-series cine-MRI. Biomech. Model. Mechanobiol. **22**(6), 1829–1846 (2023)
5. Biffi, C., et al.: 3D high-resolution cardiac segmentation reconstruction from 2D views using conditional variational autoencoders. In: International Symposium on Biomedical Imaging, pp. 1643–1646. IEEE (2019)
6. Chang, Q., et al.: DeepRecon: joint 2D cardiac segmentation and 3D volume reconstruction via a structure-specific generative method. In: International Conference on Medical Image Computing and Computer-Assisted Intervention, pp. 567–577. Springer (2022)
7. Chen, X., et al.: Shape registration with learned deformations for 3D shape reconstruction from sparse and incomplete point clouds. Med. Image Anal. **74**, 102228 (2021)
8. Chen, Y., Yang, J., Mercadier, D.S., Le, H., Fua, P.: MedTet: an online motion model for 4D heart reconstruction. arXiv preprint arXiv:2412.02589 (2024)
9. Corral-Acero, J., et al.: The 'digital twin'to enable the vision of precision cardiology. Eur. Heart J. **41**(48), 4556–4564 (2020)
10. Gillette, K., et al.: A framework for the generation of digital twins of cardiac electrophysiology from clinical 12-leads ECGs. Med. Image Anal. **71**, 102080 (2021)
11. Hu, H., Pan, N., Frangi, A.F.: Fully automatic initialization and segmentation of left and right ventricles for large-scale cardiac MRI using a deeply supervised network and 3D-ASM. Comput. Meth. Prog. Biomed. **240**, 107679 (2023)

12. Kong, F., Wilson, N., Shadden, S.: A deep-learning approach for direct whole-heart mesh reconstruction. Med. Image Anal. **74**, 102222 (2021)
13. Laumer, F., et al.: Weakly supervised inference of personalized heart meshes based on echocardiography videos. Med. Image Anal. **83**, 102653 (2023)
14. Li, L., et al.: Towards enabling cardiac digital twins of myocardial infarction using deep computational models for inverse inference. IEEE Trans. Med. Imag. (2024)
15. Niederer, S., et al.: Creation and application of virtual patient cohorts of heart models. Phil. Trans. R. Soc. A **378**(2173), 20190558 (2020)
16. Odille, F., et al.: Isotropic 3D cardiac cine MRI allows efficient sparse segmentation strategies based on 3D surface reconstruction. Magn. Reson. Med. **79**(5), 2665–2675 (2018)
17. Ukwatta, E., et al.: Myocardial infarct segmentation from magnetic resonance images for personalized modeling of cardiac electrophysiology. IEEE Trans. Med. Imaging **35**(6), 1408–1419 (2015)
18. Villard, B., Grau, V., Zacur, E.: Surface mesh reconstruction from cardiac MRI contours. J. Imag. **4**(1), 16 (2018)
19. Ye, M., Yang, D., Kanski, M., Axel, L., Metaxas, D.: Neural deformable models for 3D bi-ventricular heart shape reconstruction and modeling from 2D sparse cardiac magnetic resonance imaging. In: Proceedings of the IEEE/CVF International Conference on Computer Vision, pp. 14247–14256 (2023)
20. Yuan, X., Liu, C., Wang, Y.: 4D myocardium reconstruction with decoupled motion and shape model. In: Proceedings of the IEEE/CVF International Conference on Computer Vision, pp. 21252–21262 (2023)

Secure Medical Digital Twins: A Use-Case Driven Approach

Salmah Ahmad[1,3](✉) [iD], Bianca Bartelt[1,3] [iD], Matthias Enzmann[1,2,4] [iD],
Jörn Kohlhammer[1,2,3] [iD], Stefan Wesarg[1,2,3] [iD], and Ruben Wolf[1,4] [iD]

[1] ATHENE—National Research Center for Applied Cybersecurity,
Darmstadt, Germany
[2] Fraunhofer Cluster of Excellence Immune-Mediated Diseases CIMD,
Frankfurt am Main, Germany
[3] Fraunhofer Institute for Computer Graphics Research IGD, Fraunhoferstr. 5,
64283 Darmstadt, Germany
{salmah.ahmad,bianca.bartelt,joern.kohlhammer,
stefan.wesarg}@igd.fraunhofer.de
[4] Fraunhofer Institute for Secure Information Technology SIT, Rheinstr. 75,
64295 Darmstadt, Germany
{matthias.enzmann,ruben.wolf}@sit.fraunhofer.de

Abstract. Medical digital twins (MDTs) can support health professionals in many ways, such as in achieving a more precise diagnosis or in developing better treatment plans. For that, MDTs will need to process patients' data from multiple sources and will have to govern access to that data by health professionals from multiple institutions, and at the same time support collaboration and data sharing between them. This raises questions regarding trust and the security of processing. We illustrate two medical use cases where MDTs can have a significant impact and discuss the security scope of our MDTs.

Keywords: Digital patient twins · Data protection · Access control · Cyber security

1 Introduction

Medical digital twins (MDTs) denote a comprehensive model of a patient's history and health status. Ideally, it is based on anatomical, physiological, diagnosis- and treatment-related information from multiple sources. While often incomplete, it may also provide derived information about the patient based on known relations between the medical data. A medical digital twin also comprises information that is among the most critical and sensitive data according to the EU GDPR (General Data Protection Regulation), motivating a dedicated approach to secure such MDTs, which is further described in this paper. This approach goes beyond the current state of technology, in which mostly the transmission of medical data and the perimeter around medical data repositories are secured, while the medical data itself remains readable and exploitable.

It is essential to be able to transparently share the data contained in MDTs in a fine-grained way to give medical experts in different institutions access to certain parts of the data. The use of MDTs can be distinguished into four different groups. First, patients should be able to grant access to all data related to a specific disease for which they are treated. It should, second, also be possible to grant this access related to a treatment process across several treatment facilities. In research use cases, it is also necessary to receive access to data to build patient cohorts based on, third, certain patient features or, fourth, on organ-specific features. All use cases should be supported by easy and usable ways to grant access to specific data attributes and to use the MDTs on the medical side.

A digital twin that processes medical data of a patient requires extra care in its operation due to the sensitive nature of the data being processed. On a technical and organizational level, "extra care" translates to the implementation of measures to ensure, inter alia, authorized access, data integrity, authenticity, and confidentiality. Those are classic IT security goals which apply to the operation of a medical digital twin, too.

For every security goal, many technical solutions exist to protect the data, including access control frameworks, schemes for message authentication, digital signatures, and encryption. The challenge is to make them work together and enforce the security goals within an organization, and even more challenging beyond it, e.g., when patient data is to be shared back and forth between a physician's office and a hospital. MDTs are expected to flourish the most when medical data from different sources, like different institutions, are available to enhance the analysis of a patient's condition, to achieve a more precise diagnosis, and to allow health professionals to develop better treatment plans. We share that vision and therefore consider the flow of medical data as an important building block for an MDT. However, uncontrolled data flows may erode trust in that concept. Therefore, any solution should involve patients so that they have a say in where their medical data goes and technically enforce those decisions, e.g., by encrypting data for specific recipients. However, there can be a fine line between giving patients control over their data and overburdening them with micro-management of who gets to process which of their data. In addition, a solution should also take into account that patients may want to delegate that control to a trusted entity, e.g., a family doctor, a relative, caregiver, or a medical center. Finally, the sharing of MDTs for medical research should not be overlooked in the implementation of access control mechanisms and the access control process should support both denying and granting access in a usable way.

The contributions of our paper are the following: We provide an overview of the current research landscape on MDTs and usable interfaces for patients and medical experts. We contribute a new definition for secure MDTs that goes beyond the current perimeter- and transfer-oriented security concepts. We show new possibilities for the secure use of MDTs that are not easily feasible with traditional approaches and provide two example use cases in the areas of Parkinson's disease and inflammatory bowel disease.

2 Related Work and Background

The essential concept of a Digital Twin (DT) was first described by Grieves for manufacturing organizations [12]. In his concept, he described a physical object (PO), like a manufacturing machine, and a corresponding instance of a digital model/process (i.e., the DT) that mirrors the state of the PO using a bidirectional connection between them. The connection between the two entities should allow data to flow from the PO to the DT and vice versa. Data from the PO updates the DT's state model of the PO while data from the DT can adjust the PO's settings. However, the DT can be more than a remote control as it may process data from additional sources, e.g., data from environmental sensors, to find and propagate settings that may further improve the PO's performance. In [26] and [19], an overview of recent background and details on DTs in general can be found. Of course, DTs are not limited to manufacturing but have also been discussed for other fields like agricultural production [27], medicine [3,8,25], or smart cities [4].

A medical digital twin (MDT) can be described by observations of and measurements taken from a patient [9]. By processing a patient's data, an analytical model can be built that provides a holistic view of the patient [13] which may support health professionals in achieving timely diagnoses, possibly simulate different diagnostic scenarios, and to develop personalized treatment plans [11].

If we view an MDT as a holistic medical data model of a patient, then only a part of it will typically be required by a health professional. That part, say medical data on the patient's heart condition, can be regarded as state data which can be processed on its own, e.g., by a heart MDT. An MDT "factored out" like that can be seen as a separate DT instance (DTI) that may have connections to other DTIs, e.g., a lung MDT. Following that, a DT aggregate – comprised of DTIs – comes to mind that "allow[s] for a larger and more complete dataset regarding the operation of a type of physical object" [16]. Similarly, Haße et al. speak of a sub model of a DT when it comes to sharing parts of a DT [14]. Shared digital twins are of particular interest in the medical field because they can facilitate more effective collaboration among health professionals [25].

Information security is paramount for MDTs as a compromised MDT may have direct or indirect consequences for patients – and that includes more than their health. In its foresight report for 2030 [10], ENISA expects the "exploitation of e-health (and genetic) data" to be among the top threats by 2030 and says that such sensitive data "may be exploited or used by criminals to target individuals or by governments to control populations, e.g., using diseases [...] as a reason for discriminating against individuals". In addition, a compromised MDT may draw wrong conclusions about a patient's current state of health. In [18], general attack principles (e.g., eavesdropping) and defence mechanisms (e.g., encryption) for industrial cyber-physical systems are discussed that also apply to MDTs. Many defence mechanisms work well within an organization, like role-based access control (RBAC) [21]. However, when data leaves the control sphere of the organization, like with a shared MDT, the same or similar level of control over the data is no longer guaranteed. Data encryption is always an option for protecting shared data, but it is not a panacea, as implementing cryptography-based access control across organizations can be complex [29].

3 Medical Digital Twins

In the following, we use the term medical digital twin (MDT) in two ways. Firstly, we refer to an MDT as a collection of the medical data of some patient and secondly, to relate to the system that holds, processes, and provides access to that data (see Fig. 1).

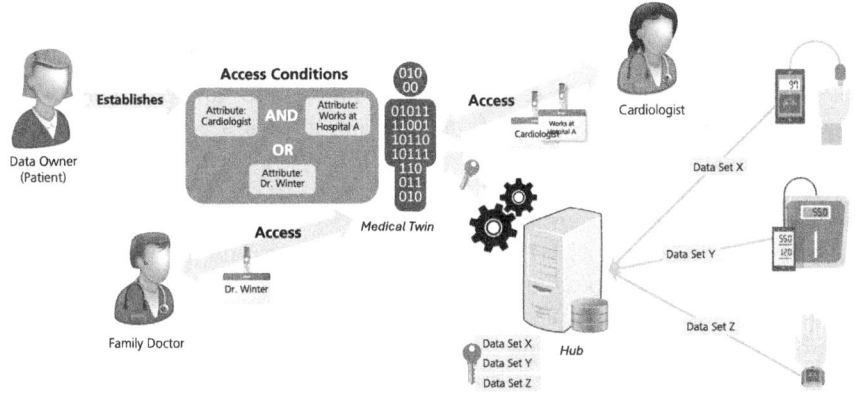

Fig. 1. Usage of a secure MDT: Data owner sets access conditions. Medical practitioners who fulfil these conditions can access the encrypted data. [17].

3.1 Shadow of a Medical Digital Twin

A digital twin can be considered as a numerical model of a physical process or an object. In case of digital patient twins, it quickly becomes obvious that a comprehensive modelling of all physiological processes going on in such an organism seems infeasible. Therefore, a medical digital twin in our concept always focuses on a subset of relevant aspects. This can be organ-driven (e.g., a digital twin of the heart), focused on a specific disease and the related therapy process, or more generally on a patient journey covering patients' pathways through treatment and the hospital.

An MDT – no matter what it specifically represents – acts as a data repository and therefore requires well-defined data structures. This allows for an aggregation of medical and health-related data. An MDT can contain information like attributes from an electronic health record, additional attributes that are relevant for medical research, sensor parameters as well as patient diary entries. For storing and accessing those data entries, data needs to be managed accordingly.

When considering specific use-cases of an MDT, a shadow [2] of the comprehensive MDT needs to be created. That means a specific subset of data structures is used for representing the case-related data, like relevant lab parameters, image-derived biomarkers, medication information, and many more. In addition, that data needs to be made accessible to the involved physicians, patients, medical nurses as well as other relevant stakeholders. Another important aspect is the integration of analysis capabilities that work on the data contained in the medical patient twin. As an example, a shadow focusing on a cardiology-related use case might integrate algorithms for detecting abnormalities in ECG data.

3.2 Collaboration, Data Sharing, and Security

A crucial aspect to be considered is that an MDT contains personal data and thus, in Europe, falls under the General Data Protection Regulation (GDPR) which governs the prerequisites, safeguards, and rules for any processing of such data. Of course, additional regulations and national laws regulating the health sector may exist but, in this paper, we focus on the security aspects of data protection. To be practically useful, an MDT needs to consider both the medical application and information security. For that, secure methods for populating the medical patient twin as well as accessing the stored data need to be provided. Our approach aims for a solution where even if an unauthorized person gains access to the MDT infrastructure the data stored therein cannot be read.

From a pure network-security point of view, MDTs may just look like another distributed system for which plenty technical and organizational measures come to mind that may be employed to achieve overall security, see for instance [18]. While an MDT is an application that heavily relies on networks, and thus surely benefits from network defence mechanisms, it is also about sharing and processing medical data. This requires more than encrypted data transmissions, where data is typically encrypted during transport only. Once the data is received, it is unclear how the data will be protected afterwards, e. g., how confidentiality and access control are implemented at the receiver's end, if at all. In our view, an MDT is about collaboration and hence, an MDT may be accessed by different user groups, e. g., the patient, health professionals from different departments and different organizations, or medical researchers. Naturally, each group will literally and technically have its own view on the MDT. In the following, we will refer to a selected portion of the detailed medical data contained in an MDT as a shadow (of the MDT).

A shadow may be plain data, data derived from an MDT, or both. From an operational and from a security perspective, a particular shadow only needs to be accessible to a certain group. For instance, if a patient is referred to a specialist, the specialist will only need access to a specific shadow of the MDT, i. e., to a limited data set. Likewise, if an MDT provides a research shadow, e. g., selected pseudonymized or anonymised patient data, medical researchers from different organizations may be given access to that data. Still, access to an MDT must only be given to qualified people, i. e., individuals that received an authorization for accessing the MDT. For instance, a specialist with access to some patient's

MDT may want to discuss the condition of the patient with a colleague and for that may want to share a shadow of the MDT with their colleague, who may or may not be working at the same practice. Likewise, a patient may see a new physician and wants to share their medical history with them. For all that, the specialist/patient needs a way to authorize access to a shadow of the MDT for another health professional. In addition, if a shadow needs to be processed by another organization, it must be securely transferred to and securely stored by that organization. Employing encrypted transmissions and encrypted storage, like full disk encryption (FDE), are a bare minimum to achieve that.

FDE does not protect individual shadows but all of them at once. At first sight, this seems to solve the problem of secure storage, but it does not. The problem is that "protecting all at once" in that case also implies "compromising all at once", which may happen when an online attacker gains access to the system that in turn has access to the encrypted storage. To that system, and hence to the attacker, the encrypted storage is "logically unencrypted", since the system automatically decrypts the stored data before it passes the data to the medical application that wants to process it. To mitigate such attacks, we aim for individually encrypted shadows of an MDT such that each (transferred) shadow is protected by its own encryption keys. Corresponding decryption keys are attached to the MDT and will only be made available after a user has demonstrated their authorization and will only be retained as long as the authorized user works with the shadow. That approach is related to the Zero Trust paradigm [5] where trust needs to be earned on every access to a resource, rather than being presumed based on weak assumptions, like "if data is accessed from within our organization's network then access is allowed". In our model, trust is earned per data set, i.e., the user must satisfy all access conditions associated with the requested data on *every* access to get the decryption key of the data.

3.3 Approach

Our approach makes use of attribute-based encryption (ABE) [23]. In ABE, more specifically in ciphertext-policy ABE, the encryptor defines an access policy in terms of attributes that a prospective decryptor must have in order to satisfy the policy, which in turn allows him to decrypt a ciphertext, i.e., the encrypted data. Each attribute is associated with a public encryption key and a secret decryption key per user. Both are issued by a so-called attribute authority that also validates if someone requesting a certain attribute is allowed to hold it. In our use cases, we expect different stakeholders to share data, and thus user groups will not necessarily be from the same organization, e.g., Hospitals A and B. For more flexibility, we use multi-authority ABE [6,20] where several attribute authorities can be established that only share an initial set of global (security) parameters. Every attribute authority may define and issue its own attributes, and the global parameters make sure that those attributes are compatible, such that they can be combined in an access policy. Larger organizations, like hospitals, may establish their own attribute authority while smaller ones, like a general practice, may employ established trust service providers.

The ABE approach is related to conventional role-based access control (RBAC) where a system-wide infrastructure controls and enforces access permissions to resources. In RBAC, resources can have security labels that determine which roles may access a certain resource (e. g., a health record), and every user in the system is assigned to roles (e. g., cardiology, administration), and roles are assigned access permissions for resources. In that sense, roles in RBAC can be compared to attributes in ABE. The major difference between RBAC and ABE is that ABE is independent of an enforcement infrastructure that controls access to data. This is advantageous if data is expected to be shared between infrastructures of different organizations. With RBAC, those infrastructures will likely have different controls, e. g., access rules and security labels. Hence, the original access conditions (roles and permissions) from the source system can no longer be relied on because they would have to be translated to the target system's, which may or may not be comparable. In contrast, access in ABE is exclusively controlled by the access policy which is an integral part of the ciphertext. And only that access policy determines the decryption key of the ciphertext and remains exactly the same, irrespective of the target system.

3.4 Usage

In Fig. 1, we have three attributes "Cardiologist", "Works at Hospital A", and "Dr. Winter" that are combined in an access policy. Although each attribute is personalized to its holder, the latter attribute is a kind of "anomaly" in ABE because it may only be held by a single person, which is why we call it a singleton. The other attributes represent groups and hence may be held by a number of persons. Personalisation of attributes is an important aspect in ABE because it prevents collusion. That is, say, two persons – one with attribute A and another with attribute B – cannot combine their attributes in order to satisfy an access policy requiring "A and B", which none of them could have satisfied alone.

We expect every encryptor to only use the most recent attributes for access policies, i. e., encryptors follow the authorities' public key updates, which are accessible through a public directory. If a user is no longer eligible for a certain attribute, he is not permitted to renew his secret attribute key with the issuing authority. Renewal of a secret key becomes necessary whenever an attribute's corresponding public key is updated. Without renewal, a user will not be able to decrypt ciphertexts that were encrypted with recently updated public attribute keys. The time frame for public-key updates is up to the authority/organization and may depend on the organization's risk tolerance. For instance, a risk-averse organization may update an attribute key whenever the group of users is reduced, e. g., when someone leaves the organization. Others with more risk appetite (or possibly less sensitive data) may wait until a certain threshold of changes is reached, or some fixed time frame has passed, before they update an attribute key. Note that ciphertexts encrypted under some access policy will not change if the public key of an attribute from the policy is updated, i. e., someone with "old" keys can still decrypt them. For that, our current approach is to have lazy updates of key encryption keys (KEKs = ABE keys). That is, we use hybrid encryption, where data encryption keys (DEKs) are used to encrypt the "actual data" and KEKs to encrypt the DEKs. Now, "lazy" means that ciphertexts of DEKs are updated to the current ABE keys with the next write operation, i. e., DEKs are re-encrypted using current KEKs. Another approach would be to update a DEK ciphertext whenever an attribute of its access policy changes. So again, this is a matter of risk tolerance, but also of work load because existing DEK ciphertexts would have to be constantly re-encrypted in the background.

Still, if an attacker manages to break into the system of an authorized user that is currently working with a shadow, the attacker may learn what is contained in that shadow. However, the attacker will not gain access to all other shadows because their decryption keys are unavailable without separate authorizations. The challenge of the cryptographic approach is that keys on several access levels need to be generated and efficiently managed to provide correct handling for the different stakeholders.

A final requirement is the necessity to integrate population norms into the MDT. This allows to compare individual patient parameters to those that can be expected in a healthy person or to those specific for the staging of a disease. The other direction is the derivation of cohorts when aggregating the information stored in several shadows. This requires a partial access to patient-individual parameters that needs to be granted to researchers for verifying hypotheses or preparing patient studies.

4 Use Cases

Medical digital twins (MDTs) have vast potential in the field of healthcare, offering innovative solutions for the treatment of various medical conditions. In the following sections, we explore two use cases where MDTs can have a significant impact: neurodegenerative diseases and inflammatory bowel disease.

4.1 Neurodegenerative Diseases

Increasing life expectancy is a key driver of the growing incidence of neurodegenerative diseases (NDDs) such as Alzheimer's, Parkinson's and ALS [24]. These diseases lead to a progressive loss of cognitive and motor functions, which severely affects quality of life. As NDDs cannot yet be cured, physicians focus on relieving the symptoms, which is usually very complex, time-consuming and cost-intensive [28]. The integration of secure MDTs, as defined above, offers a promising approach to address these challenges.

In the context of NDDs, our MDT serves as both a structured, disease-specific repository of a patient's health data and the system that manages, processes, and provides secure access to this data. The MDT captures and organizes heterogeneous, multimodal health data most relevant to NDDs, such as longitudinal cognitive assessments, motor function tests, medication records, patient diary entries and neuroimaging data (e. g., MRI, PET, SPECT scans). For example, the MDT can be continuously updated with imaging results that reveal disease-specific changes, such as dopaminergic neuron loss in Parkinson's or amyloid plaques in Alzheimer's. This enables physicians to monitor disease progression, detect early biomarkers and personalize treatment strategies.

Access to the MDT is strictly controlled, as outlined in our approach. Only authorized stakeholders, such as neurologists, caregivers or family members, can access specific shadows of the MDT. For example, a neurologist may access longitudinal MRI scans and cognitive test results to monitor disease progression, while a caregiver may only view medication schedules and daily activity logs. As NDDs progress and the patient may no longer be able to make health decisions for themselves, the MDT allows for secure delegation of access to trusted family members or legal guardians. This ensures continuity of care while maintaining strict data protection. Building on these capabilities, our proposed MDT securely brings together key NDD data, enabling personalized care, clear updates for patients and caregivers and ultimately contributes to an overall improvement in the treatment of NDDs.

4.2 Inflammatory Bowel Disease

Inflammatory bowel disease (IBD) is a chronic condition that is costly to both patients and the healthcare system. It is characterized by relapsing-remitting symptoms, which cause inflammation of the gastro-intestinal tract [1,22]. Well-known varieties of IBD are Crohn's disease or ulcerative colitis. Symptoms include diarrhea, abdominal pain, and, in the case of ulcerative colitis, rectal bleeding. The progression of the disease frequently extends over several decades, and its prevalence is increasing at a global scale [7].

Managing this condition is often challenging, primarily due to limited access to comprehensive patient records. This restricts informed decision-making and can result in a significant amount of manual data processing if the complete medical history of a patient is to be considered. By helping physicians analyze the patient's medical history, symptoms, and response to previous treatments, MDTs can support insights that enable personalized care. Treatment planning is also informed by medical imaging results. Endoscopies are a diagnostic tool that can provide more qualitative information about the location and extent of intestinal inflammation. Echographies, on the other hand, are used to assess disease severity, such as in the *Limberg score*. The availability of these results within the MDT facilitates the identification of correlations with other components of the data set. This comprehensive monitoring can assist in detecting early indications of disease flare-ups or complications, enabling timely intervention and averting severe consequences.

Moreover, patients have the capacity to record symptoms themselves within the MDT, thereby becoming active participants in their own care by monitoring their health parameters. As IBD is a lifelong disease, being able to easily and quickly record daily symptoms is very helpful, as it provides real-time data on various parameters such as inflammation levels, bowel movements and medication adherence. Within our secure MDT approach, the management of patient data pertaining to IBD (including symptoms, laboratory results and imaging biomarkers) is accomplished through individually encrypting each data entry. This ensures that only authorised users are able to decrypt the precise information they require.

Overall, our MDT approach transforms IBD management by combining continuous patient-reported data, endoscopic and imaging results, and laboratory biomarkers in a secure, collaborative framework – supporting personalized treatment insights, reducing manual record reconciliation, and safeguarding patient trust through robust cryptographic controls.

5 Future Work

Further research and development are necessary to implement our proposed medical digital twin (MDT) in personalized healthcare. Our goal is to establish a solid foundation for the creation and operation of MDTs. This involves collecting medical data and diagnosis- and treatment-related information from trusted sources to create comprehensive models of a patient's history and health status. Throughout the development process, we maintain close contact with potential users from both the medical community and patients to ensure a user-centered approach. We are currently conducting a survey aimed at practicing physicians to better understand the current landscape of medical data exchange. In close collaboration with medical experts, we will further develop practical use cases. This will include the development of user-friendly interfaces that allow medical experts to request access to and utilize specific data within the MDT. Medical personnel will be able to populate the MDT with new examination results. On the other side, patients will be able to intuitively grant, revoke or change access rights to specific data while also being able to integrate personal data into the MDT. For that, we will implement encryption-based access concepts to ensure the security of processing. We are working on interoperability with the FHIR standard and HL7 [15]. The newest version of FHIR (v6.0.0), which will probably be released in the coming months, will support an attribute-level rights management, which may allow us to integrate our encryption concept with the HL7/FHIR approach.

6 Conclusion

This paper introduces the concept of secure medical digital twins as a new means to support important medical use cases that require the sharing of medical data. By securing the data and the access to the data in a decentralized way, we foster the secure sharing of medical data which can otherwise be cumbersome with the more traditional perimeter- and transfer-oriented security concepts. This has benefits for patients, who have full transparency of the grants of access to their medical data, and for medical experts and institutions, who can rely on these patient approvals during the treatment process and for research purposes. We showed in two example use cases how these benefits manifest themselves during treatment planning for Parkinson's disease and for inflammatory bowel disease. We are working on several prototypical implementations that will provide the basis for future medical information systems as well as tele-medical applications.

Acknowledgments. This research work was supported by the National Research Center for Applied Cybersecurity ATHENE. ATHENE is funded jointly by the German Federal Ministry of Research, Technology and Space and the Hessian Ministry of Science and Research, Arts and Culture. The work was partly funded by the Fraunhofer Cluster of Excellence Immune-mediated Diseases CIMD.

Disclosure of Interests. The authors have no competing interests to declare that are relevant to the content of this article.

References

1. Baumgart, D.C., Carding, S.R.: Inflammatory bowel disease: cause and immunobiology, pp. 1627–1640. Elsevier (2007)
2. Becker, F., et al.: A conceptual model for digital shadows in industry and its application. In: Ghose, A., Horkoff, J., Silva Souza, V.E., Parsons, J., Evermann, J. (eds.) ER 2021. LNCS, vol. 13011, pp. 271–281. Springer, Cham (2021). https://doi.org/10.1007/978-3-030-89022-3_22
3. Björnsson, B., Borrebaeck, C., Elander, N., Gasslander, T., Gawel, D.R., et al.: Digital twins to personalize medicine. Genome Med. **12**(4) (2020)
4. Bujari, A., Calvio, A., Foschini, L., Sabbioni, A., Corradi, A.: IPPODAMO: a digital twin support for smart cities facility management. In: GoodIT '21, pp. 49–54. ACM (2021)
5. Bundesamt für Sicherheit in der Informationstechnik (BSI): Positionspapier Zero Trust 2023. Technical report, Bundesamt für Sicherheit in der Informationstechnik (BSI), Bonn (2023)
6. Chase, M.: Multi-authority attribute based encryption. In: Vadhan, S.P. (ed.) TCC 2007. LNCS, vol. 4392, pp. 515–534. Springer, Heidelberg (2007). https://doi.org/10.1007/978-3-540-70936-7_28
7. Collaborators, G.B.D., et al.: Global, regional, and national incidence, prevalence, and years lived with disability for 354 diseases and injuries for 195 countries and territories, 1990-2017: a systematic analysis for the Global Burden of Disease Study 2017. University of Leicester (2018)
8. Coorey, G., et al.: The health digital twin to tackle cardiovascular disease - a review of an emerging interdisciplinary field. NPJ Digital Med. **5**(126) (2022)
9. Erol, T., Mendi, A.F., Doğan, D.: The digital twin revolution in healthcare. In: Proceedings of 4th International Symposium on Multidisciplinary Studies and Innovative Technologies (ISMSIT), pp. 1–7. IEEE (2020)
10. European Union Agency for Cybersecurity (ENISA): Identifying Emerging Cyber Security Threats and Challenges for 2030 (2023)
11. Farsi, M., Daneshkhah, A., Hosseinian-Far, A., Jahankhani, H. (eds.): Digital Twin Technologies and Smart Cities. IT, Springer, Cham (2020). https://doi.org/10.1007/978-3-030-18732-3
12. Grieves, M.W.: Product lifecycle management: the new paradigm for enterprises. Int. J. Product Dev. **2**(1), 2 (2005)

13. Haleem, A., Javaid, M., Singh, R.P., Suman, R.: Exploring the revolution in healthcare systems through the applications of digital twin technology. Biomed. Technol. **4**, 28–38 (2023)
14. Haße, H., Valk, H., Möller, F., Otto, B.: Design principles for shared digital twins in distributed systems. Bus. Inf. Syst. Eng. **64**(6), 751–772 (2022)
15. Health Level Seven International, Inc.: HL7 FHIR (2025). https://hl7.org/fhir/. Accessed 24 Jul 2025
16. Human, C., Basson, A.H., Kruger, K.: A design framework for a system of digital twins and services. Comput. Ind. **144** (2023)
17. Icons: DB server by OpenClipArt (CC0 1.0); Free Large Boss Icon Set by www.aha-soft.com; Health device icons designed by macrovector_official / Freepik
18. Jiang, Y., Wu, S.M.R., Liu, M., Luo, H., Kaynak, O.: Monitoring and defense of industrial cyber-physical systems under typical attacks: from a systems and control perspective. IEEE Trans. Ind. Cyber-Phys. Syst. **1** (2023)
19. Jones, D., Snider, C., Nassehi, A., Yon, J., Hicks, B.: Characterising the digital twin: a systematic literature review. CIRP J. Manuf. Sci. Technol. **29**, 36–52 (2020)
20. Lewko, A., Waters, B.: Decentralizing attribute-based encryption. In: Paterson, K.G. (ed.) EUROCRYPT 2011. LNCS, vol. 6632, pp. 568–588. Springer, Heidelberg (2011). https://doi.org/10.1007/978-3-642-20465-4_31
21. National Institute of Standards and Technology (NIST): American National Standard for Information Technology - Role Based Access Control, INCITS 359-2012 (R2022) (2012)
22. Piovani, D., Danese, S., Peyrin-Biroulet, L., Bonovas, S.: Inflammatory bowel disease: estimates from the global burden of disease 2017 study. Wiley Online Library, pp. 261–270 (2020)
23. Sahai, A., Waters, B.: Fuzzy identity-based encryption. In: Cramer, R. (ed.) EUROCRYPT 2005. LNCS, vol. 3494, pp. 457–473. Springer, Heidelberg (2005). https://doi.org/10.1007/11426639_27
24. Su, D., et al.: Projections for prevalence of Parkinson's disease and its driving factors in 195 countries and territories to 2050: modelling study of global burden of disease study 2021. BMJ **388** (2025)
25. Vallée, A.: Digital twin for healthcare systems. Front. Digital Health **5** (2023)
26. VanDerHorn, E., Mahadevan, S.: Digital twin: generalization, characterization and implementation. Decision Support Syst. **145** (2021)
27. Verdouw, C., Tekinerdogan, B., Beulens, A., Wolfert, S.: Digital twins in smart farming. Agric. Syst. **189** (2021)
28. Zahra, W., et al.: The global economic impact of neurodegenerative diseases: opportunities and challenges. In: Keswani, C. (ed.) Bioeconomy for Sustainable Development, pp. 333–345. Springer, Singapore (2020). https://doi.org/10.1007/978-981-13-9431-7_17
29. Zhang, L., Kan, H., Huang, H.: Patient-centered cross-enterprise document sharing and dynamic consent framework using consortium blockchain and ciphertext-policy attribute-based encryption. In: CF '22, Proceedings of the 19th ACM International Conference on Computing Frontiers, 2022. pp. 58–66. ACM (2022)

Explainable Prediction of Recurrence After Prostate Cancer Radiotherapy Using *in Silico* digital twin model and machine learning

Valentin Septiers[1,2(✉)], Carlos Sosa-Marrero[1], Eleonora Poeta[3], Hilda Chourak[1], Aurélien Briens[1], Renaud De Crevoisier[1], Maria A. Zuluaga[2], and Oscar Acosta[1]

[1] Univ Rennes - CLCC Eugène Marquis INSERM - LTSI - UMR 1099, 35000 Rennes, France
valentin.septiers@univ-rennes.fr
[2] Data Science Department, EURECOM, 06410 Biot, France
[3] Politecnico di Torino, Turin, Italy

Abstract. Biochemical recurrence (BCR) for prostate cancer (PCa) patients treated with External Beam Radiation Therapy (RT) has an incidence rate of up to 20 %. Thus, predicting BCR after PCa RT appears crucial for personalising treatments. Current approaches, such as radiomics and deep learning, applied to clinical and *in vivo* imaging data, suffer from limited explainability. This paper introduces a pipeline for predicting BCR by integrating clinical data with biologically grounded features derived from in silico digital twin simulations, supported by two explainability analyses. Specifically, we leverage a previously developed *in silico* digital twin model to simulate tumour growth and response to radiation for 315 PCa patients retrospectively treated with RT. A logistic regression model was identified as the best predictor, integrating clinical characteristics and biologically interpretable features extracted from simulations (AUC = 0.73). To enhance explainability, a local perturbation analysis is performed to quantify the influence of individual radiobiological parameters within the *in silico* model. Additionally, SHapley Additive exPlanations (SHAP) were applied to evaluate the contribution of each feature to the BCR prediction. By linking simulation-driven parameter importance with feature-level explanations, the pipeline provides coherent insights at the mechanistic and statistical levels.

Keywords: prostate cancer · explainability · digital twin · recurrence prediction

1 Introduction

Prostate cancer (Pca) is the second most diagnosed cancer in men in the world and the fifth leading cause of death [3]. External Beam Radiation Therapy (RT)

is the clinical standard treatment for localised PCa administered to 60 % of patients [14], which allows for control of the tumour in the majority of cases. However, biochemical recurrence (BCR), may occur in 1.8–5.2 %, 7.9–11.8 % and 12.0–17.2 % of patients with respectively low, intermediate and high-risk tumours within 8 years after the treatment [25]. Thus, it is crucial to assess patient risk by predicting BCR after PCa RT to personalise treatments afterwards.

Classical BCR prediction models are based on tumour control probability [4] describing the dose-effect relationship on a population without taking into account the heterogeneous nature of tumours. Radiomics models introduced the extraction of numerous image-based features, particularly from Magnetic Resonance Imaging (MRI) or Positron Emission Tomography (PET) [7,18]. Most recently, deep learning models have emerged as appealing tools for prediction, yielding better results than radiomics [17]. These data-driven approaches are based on clinical and/or imaging features for the prediction, achieving better performance but requiring large-scale training, being highly dependent on data quality, and opertaing as *black-box* methods [6,10]. While recent studies have employed explainability techniques such as SHapley Additive exPlanations (SHAP) [18,24], the resulting explanations are often difficult to interpret due to the abstract nature of the input features. In particular, commonly used descriptors such as texture or wavelet-based features, although identified as influential, lack direct biological or clinical interpretability [6].

Computational multiscale *in silico* simulation models allow the simulation of RT on multiple virtual tumours or digital twins, providing insights into tumour evolution during RT and demonstrating comparable or better results compared to data-driven approaches [19]. These models are based on the integration of several biological mechanisms related to the 5 R's of radiobiology: Reoxygenation, Repopulation, DNA Repair, Radiosensitivity, and Redistribution in the cell cycle [11]. Given their nature, they can be used to better understand the response of patients to RT by extracting interpretable biological *in silico* features [19]. While promising, explainability techniques have not been applied to predictions from *in silico* simulation models.

In this study, we propose a pipeline for predicting BCR after PCa RT by combining clinical information and a computational *in silico* digital twin model with machine learning (ML) techniques. The pipeline extracts biologically meaningful features, offering insights into tumour dynamics and treatment response. To our knowledge, this study is the first to include direct or clinically interpretable biological characteristics to predict BCR, supported by explainability techniques to better understand how each characteristic influences tumour progression and BCR. We leverage an *in silico* model, based on [22], to simulate RT for a cohort of 315 patients and extract features used in SHAP-based explainability analysis. In parallel, a local perturbation analysis (LPA) is performed to identify the key radiobiological parameters influencing tumour evolution. Together, these approaches provide a comprehensive and explainable framework for BCR prediction.

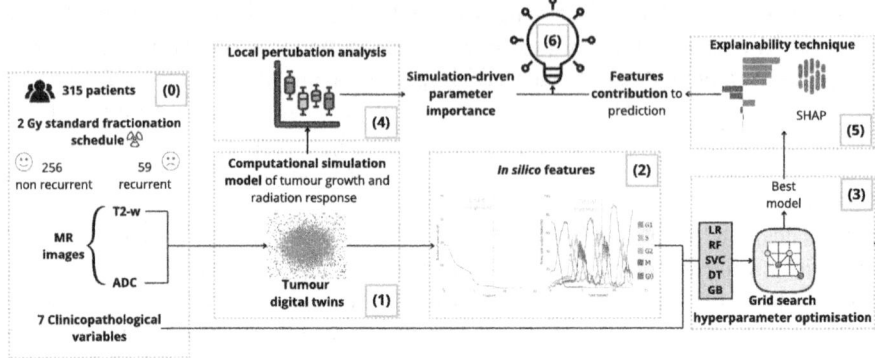

Fig. 1. Workflow. Patient dataset with 315 patients (59 BCR) treated with 2 Gy standard fractionation schedule. Two MR image sequences and 7 clinical variables are available (0); Tumour digital twins are created from MR images for each patient and computational *in silico* simulations of the prescribed treatment are performed (1); Extraction of *in silico* features from the simulations (2); Grid search optimisation to predict BCR and select a best ML algorithm according to the AUC (3); LPA of the *in silico* model to assess simulation-driven parameter importance (4); SHAP explainability assessment of feature contribution to the prediction of BCR (5); Association of simulation-driven parameter and feature importance (6).

2 Materials and Methods

Figure 1 illustrates the proposed pipeline. After patient dataset curation (Step (0)), a tumour digital twin model is built, then computational simulations are performed (Step (1)) from which *in silico* features are extracted (Step (2)). A grid search optimisation is performed to predict BCR (Step (3)), followed by a LPA of the *in silico* model (Step (4)). The results of a SHAP analysis on the best predictor (Step (5)) are associated with the outcomes of LPA (Step (6)).

2.1 Patient Dataset (Step (0))

A retrospective cohort of 315 patients with localised PCa, following the principles of the Declaration of Helsinki, is used. The patients underwent conventional RT: 74–80 Gy in 37–40 fractions (2 Gy per fraction) (see Table 1). The available clinical variables include: age at treatment initiation, TNM classification, Gleason score, pre-treatment PSA level, ISUP grade group [8], total RT dose administered, and hormonal treatment status (with or without hormonal therapy).

Prior to treatment, 3T multiparametric MRI scans were acquired, including T2-weighted (T2w) and diffusion-weighted sequences with multiple b-values. Detailed MRI acquisition parameters are described in [10]. T2w images were pre-processed with N4 bias field correction, and Apparent Diffusion Coefficient (ADC) maps were computed from diffusion-weighted images. Prostate and tumour regions were manually segmented by expert radiation oncologists on

Table 1. Patients description and tumour characteristics

Patients description	Tumour characteristics			
Number of patients	Pre-treatment PSA (ng/mL)			
315	$PSA \leq 7$	$7 < PSA \leq 11$	$11 < PSA \leq 20$	$PSA > 20$
Median age (years)	27 %	20 %	32 %	21 %
71	Clinical stage (T stage)			
Recurrence	T1	T2 - T3	T4	T5
NonBCR / BCR	13 %	68 %	10 %	9 %
256 / 59	Gleason score			
Total Dose (Gray)	6	7	8	9
74 - 80	17 %	61 %	13 %	9 %

T2w images, and the contours were then propagated to the co-registered ADC images.

Patients were monitored through clinical examinations and PSA testing every six months for five years following RT. Based on the Phoenix criteria, 59 out of 315 patients experienced BCR [1].

2.2 Computational *in Silico* digital twin model (Step 1)

A previously developed multiscale computational *in silico* digital twin model simulating tumour growth and response to RT [22] is employed. It allows the simulation of both 2D and 3D tumours evolution. This model includes the major radiobiological mechanisms, occurring at various temporal and spatial dimensions (Angiogenesis, division of healthy and tumour cells, oxygenation, and response to irradiation of tumour, healthy, and endothelial cells). The mechanisms of angiogenesis and oxygenation are simulated with two distinct reaction-diffusion equations. The division of cells is defined according to the cell cycle phases (G_1 (Gap 1), S (Synthesis), G_2 (Gap 2), M (Mitosis) and G_0 (Quiescence)) and duration. The response to irradiation is based on the computation of the survival fraction with the Linear Quadratic Model ([14]), function of the oxygen concentration in the cell/tissue. All these mechanisms and equations provide several radiobiological parameters difficult to adapt specifically to each patient, thus, for this study, they are set based on prior literature [22]. Furthermore, this *in silico* model was implemented in C++, based on the Multiformalism Modeling and Simulation Library (M2SL) [12], allowing the integration of these mechanisms arising at various temporal and spatial scales. Practically, it considers digital twins of patient tumours, placed in a grid, in which each pixel represents a cell. Each cell is considered evolving with its own parameters, attributes, and state regarding the radiobiological environment. Solving reaction diffusion equations, calculating the survival fraction, and tracking cell evolution at each time step, allows to simulate the complete evolution of the tumour in response to the treatment on the grid. All the details about this *in silico* model are fully described in [22].

2.3 Digital Twins and Patient-Specific *in Silico* simulations (Step 2)

2D digital tissues representing patient-specific tumours are built from MR imaging data. Volume and sphericity features are extracted from T2-weighted images within the tumour region using the Pyradiomics library [23], after resampling all images to a resolution of 2 mm × 2 mm × 2 mm with B-spline interpolation. Tumour density is estimated from the mean ADC value inside the tumour using a linear transformation [15]. The simulation grid is defined with a cell pixel size of 20.0 μm, corresponding to the average diameter of a cell. Vascular density is set at 3.8 % to model a poorly vascularised tumour core.

The conventional RT treatment prescribed to each patient is simulated up to four weeks post-treatment. The resulting *in silico* simulations produce a set of biologically interpretable features, offering insights into tumour dynamics during and after therapy. These include tumour volume and density, the percentage of tumour cells in each phase of the cell cycle, the proportion of killed cells, and the volume of undamaged tumour cells.

2.4 Prediction of BCR (Step 3)

Highly correlated features (Spearman coefficient $\geq$ 0.8) between clinical and extracted *in silico* variables are removed. Then, feature selection is performed using affinity propagation [9]. Multiple ML algorithms are evaluated for BCR prediction, including logistic regression (LR), random forest (RF), support vector classifier (SVC), decision tree (DT), and gradient boosting (GB). A hyperparameter grid search is conducted to identify the best-performing model. Due to class imbalance towards no-BCR label (19% of patients with BCR), Synthetic Minority Over-sampling TEchnique (SMOTE) is applied to augment the minority class [5]. The choice of the best ML algorithms and hyperparameters is made according to the best Area Under the ROC Curve (AUC) found to predict BCR. Then, an ablation study is conducted on the best ML algorithms to assess the performance of models trained on clinical features alone, *in silico* features alone, or a combination of both. For each prediction model, a cross-validation stratified 10-fold is employed with 100 repetitions. Additional comparisons are made with models incorporating radiomics and clinical + radiomics features. Model performance is evaluated using AUC, Accuracy (ACC), Precision (PREC), Recall (REC), and F1 score (F1).

2.5 Explainability Assessment (Steps 4-6)

To assess the importsance of local parameters in the *in silico* model, a Local Perturbation Analysis (LPA) is performed (Step (4) in Fig. 1). Twenty-one digital tissues are generated from histopathological data extracted from radical prostatectomy specimens. For each radiobiological parameter, a range of values, defined based on a prior study [22], is evaluated across all digital tissues. A non-dimensional Relative Importance Coefficient (RIC) is computed, adapted from classical sensitivity analysis techniques [2], and defined as the fraction of change in a parameter that propagates to a change in the output of the *in silico* model. This perturbation-based strategy is conceptually aligned with local explanation techniques widely used in Explainable AI, such as LIME [20] and input perturbation methods for feature attribution [13,21]. It allows quantifying the influence of each input parameter by observing the corresponding variation in the simulation outcome. For this analysis, the integral of tumour density over the course of RT is used as the output function, as it captures tumour evolution during treatment. By systematically varying one parameter at a time and monitoring the output response, this analysis provides biologically meaningful insights into the behaviour of the *in silico* model and supports its interpretability in the context of treatment response prediction.

To assess the contribution of each feature to the prediction of biochemical recurrence (BCR), both local and global SHAP (SHapley Additive exPlanations) analyses are conducted for the best-performing model that combines clinical and *in silico* features (Step (5) in Fig. 1) [16]. To further investigate model explainability, an explainability-by-removal analysis is performed by sequentially excluding the top- and least-contributing features identified through global SHAP values, one at a time, and evaluating the impact on predictive performance. Finally, the feature contributions obtained through SHAP are compared with the parameter importances derived from the *in silico* LPA. This association enables the interpretation of predictions in terms of both simulation dynamics and model behaviour, offering a comprehensive view of the predictive pipeline (Step (6) in Fig. 1).

3 Results and Discussion

Table 2. BCR prediction performance for each combination of features. The model 3 is the proposed approach combining clinical + *in silico* features. Bold denotes top results. PREC, REC and F1 are computed for class 0 (no_BCR) and class 1 (BCR). Features 1^+ and 2^+ are the top contributors to the prediction according to global SHAP analysis.

Models	AUC	ACC	PREC		REC		F1	
			0	1	0	1	0	1
Model 1 : Clinical	0.68	**0.71**	0.87	0.28	**0.76**	0.47	**0.79**	0.33
Model 2 : *In silico*	0.67	0.6	0.89	0.28	0.57	**0.71**	0.67	0.39
Model 3 : Clinical + *in silico*	**0.73**	0.69	**0.91**	**0.34**	0.69	0.67	0.76	**0.43**
Model 4 : Radiomics	0.55	0.63	0.83	0.21	0.69	0.38	0.74	0.25
Model 5 : Clinical + radiomics	0.65	0.67	0.87	0.28	0.7	0.53	0.76	0.34
Explainability by removal								
Model 3 without feature 1^+	0.65	0.67	0.89	0.32	0.68	0.62	0.75	0.4
Model 3 without feature 2^+	0.63	0.57	0.89	0.28	0.55	0.71	0.65	0.38

3.1 Prediction of BCR

Affinity propagation clustering on the combined clinical and *in silico* feature set selects three clinical variables (Gleason score, pre-treatment PSA level, and hormonal therapy status) and 5 *in silico* features: (1) the ratio of undamaged tumour cell volume between the 4th and 2nd weeks of treatment (4_2weeksRattumNotDamVol), (2) the percentage of tumour cells in the G1 phase at week 8 (8weeksG1Phase), (3) the amplitude of the cosine function fitted to the percentage of cells in M phase (cosFitParamAMPhase), (4) the trend of the percentage of cells in S phase between weeks 10 and 12 (curve-Trend10_12weeksSPhase), and (5) the period of the cosine fit for the G1 phase (cosFitParamTauG1Phase).

Table 2 reports performance metrics for all feature combinations. The best-performing model identified through grid search (model 3) is a LR classifier (parameters: C = 0.1, max_iter = 100000, penalty = 'l1', solver = 'liblinear', and random_state = 42) trained on the selected clinical and *in silico* features.

3.2 Explainability Analysis

Figure 2 presents the results of LPA applied to the parameters of the *in silico* model. The computed RIC is reported for all radiobiological parameters, with colours indicating the five underlying biological mechanisms. Parameters related to angiogenesis (red), healthy cell division (orange), and the irradiation response of healthy and endothelial cells (purple) show negligible importance based on their RIC values. In contrast, parameters associated with tumour oxygenation (green) and tumour cell proliferation (light blue) demonstrate the highest influence on the model output. Parameters describing the tumour's response to irradiation (dark blue), which depends on oxygenation, also contribute significantly. Among the most influential parameters, the oxygen diffusion coefficient (D^{O_2}), the pO_2 level of pre-existing endothelial cells (pO_2^{preEnd}), and the tumour cell doubling time (T_{tum}) exhibit a locally positive effect on the integral of tumour density over treatment. Conversely, the maximum oxygen consumption rate ($V_{max}^{O_2}$) and the necrosis threshold pO_2 (pO_2^{nec}) show a locally negative contribution. This analysis provides mechanistic insights into how tumour biology shapes simulation outcomes.

Figure 3 presents the global SHAP analysis and Fig. 4 the local SHAP explanation for one BCR patient correctly classified by model 3. SHAP values for the local analysis are computed with respect to the prediction for the class of interest (BCR). In this example, the Gleason score is the most influential feature, strongly pushing the prediction toward the BCR class. Conversely, several features have near-zero SHAP values, indicating little to no contribution to the prediction, consistent with the global SHAP analysis.

Explainability-by-removal results are reported in Table 2 for the most important features identified by global SHAP. The two most contributing features,

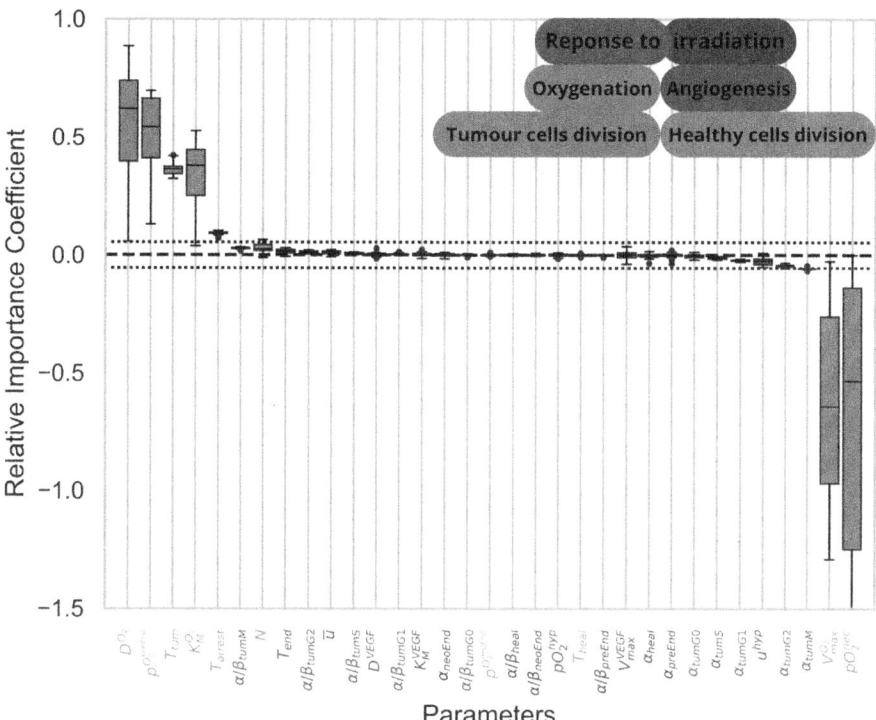

Fig. 2. Results of the LPA. RIC coefficient is shown for each radiobiological parameter. Colours correspond to each radiobiological mechanisms on the top right. The standard deviation of each boxplot shows the variability among all 21 digital tissues. The standard deviation of the last parameter shown was too high, so the boxplot is cropped for better visualisation. The 2 dotted lines around 0 correspond to $+/-0.05$ representing the separation between parameters of no importance and important ones.

CosFitParamAMPhase (*in silico*) and gleason_score (clinical), are labeled as 1^+ and 2^+, respectively. Conversely, the least contributing features, CosFitParamTauG1Phase and curveTrend10_12weeksSPhase (both *in silico*), are denoted 1^- and 2^-. Removing feature 1^+ results in a slight performance drop, suggesting that its contribution, though high, is partially compensated by other features. In contrast, removing feature 2^+ leads to a substantial drop in performance particularly in accuracy (ACC), recall (REC), and F1 score for the no_BCR class, highlighting its critical predictive role. Removing 1^- or 2^- does not significantly affect performance, confirming their limited influence.

The radiobiological *in silico* features used for prediction offer insights into tumour evolution during and after RT, complementing clinical variables. Most of these features are linked to the cell cycle, particularly the G1, S, and M phases, reflecting tumour cell proliferation. This aligns with the results of the LPA, which identified tumour cell division as a key mechanism. Additionally,

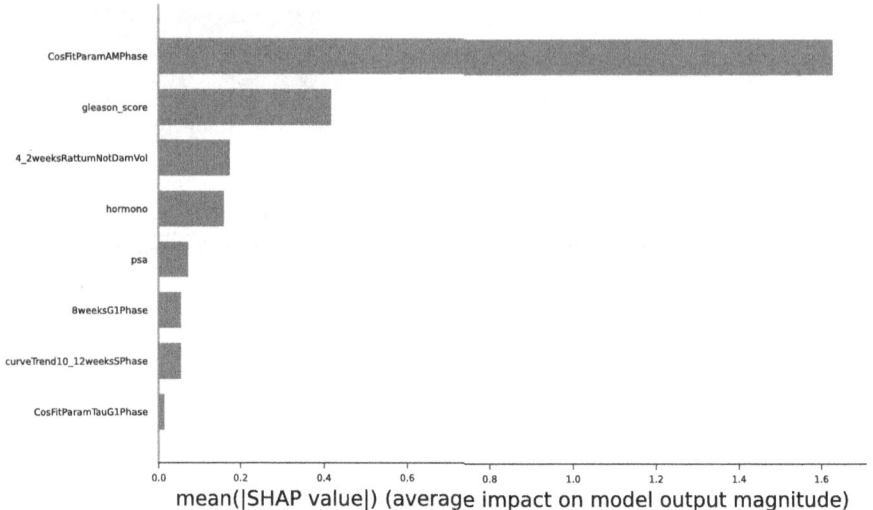

Fig. 3. Global SHAP analysis. The mean SHAP value for each feature of prediction model 3 is presented, showing the average impact on model output.

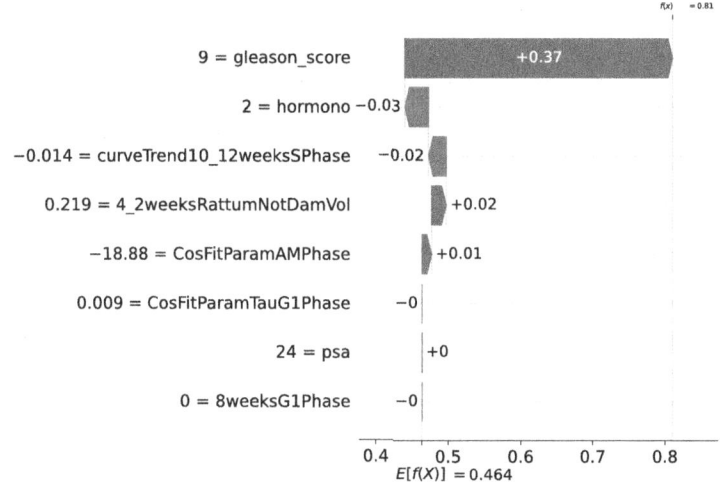

Fig. 4. Local SHAP analysis for one BCR patient, well predicted by model 3. The local SHAP value for each feature of prediction model 3 is presented, showing how each feature influences the prediction towards the BCR class.

one *in silico* feature reflects the volume of undamaged tumour cells and relates to the tumour's response to irradiation. This is coherent with LPA findings that emphasise the importance of oxygenation-linked radiation response mechanisms.

4 Conclusions

This study presents a novel integration of explainable AI techniques into a predictive pipeline for BCR following PCa RT. By combining clinical data with biologically grounded features derived from a computational *in silico* digital twin model, the approach captures tumour dynamics and treatment response with enhanced understandability and predictive power. The resulting LR model achieves an AUC of 0.73, outperforming models based solely on clinical or radiomics data. To explain model behavior, SHAP values and LPA are employed, bridging feature-level and mechanistic understanding. The proposed framework not only enables accurate BCR prediction but also lays the groundwork for explainable risk-based patient stratification and biologically informed, personalised treatment planning in PCa care.

This study has limitations that will be explored in future work. Firstly, the computational *in silico* digital twin model must be enriched with additional radiobiological mechanisms, such as the immune response. The 3D simulations are being explored, taking into account the segmented mesh of the tumor to better visualise tumor shrinkage and evolution, providing visual explanations. Other types of images, such as PET, could provide more information on the vascularisation of the tumour to build more personalised digital twins. The LPA offered explanations on local parameter effects. A global analysis, including the computation of dependencies between parameters, could be performed to assess the links between all the parameters. Then, other feature selection techniques or ML algorithms may have been applied and tested.

Acknowledgments. With financial support from ITMO Cancer of Aviesan within the framework of the 2021–2030 Cancer Control Strategy, on funds administered by Inserm.

Disclosure of Interests. The authors have no competing interests to declare that are relevant to the content of this article.

References

1. Abramowitz, M.C., et al.: The phoenix definition of biochemical failure predicts for overall survival in patients with prostate cancer. In: Cancer (2008)
2. Beven, K.: A sensitivity analysis of the penman-monteith actual evapotranspiration estimates. In: J. Hydrol. (1979)
3. Bray, F., et al.: Global cancer statistics 2022: GLOBOCAN estimates of incidence and mortality worldwide for 36 cancers in 185 countries. In: CA. Cancer J. Clin. (2024)
4. Chanrion, M.-A., et al.: The influence of the local effect model parameters on the prediction of the tumor control probability for prostate cancer. In: Phys. Med. Biol. (2014)

5. Chawla, N.V., et al.: SMOTE: synthetic minority over-sampling technique. In: J. Artif. Intell. Res. (2002)
6. Duenweg, S.R., et al.: T2-weighted MRI radiomic features predict prostate cancer presence and eventual biochemical recurrence. In: Cancers
7. Dutta, A., et al.: Robustness of magnetic resonance imaging and positron emission tomography radiomic features in prostate cancer: impact on recurrence prediction after radiation therapy. In: Phys. Imaging Radiat. Oncol. (2024)
8. Epstein, J.I., et al.: The 2014 international society of urological pathology (ISUP) consensus conference on gleason grading of prostatic carcinoma: definition of grading patterns and proposal for a new grading system. In: Am. J. Surg. Pathol. (2016)
9. Frey, B.J., Dueck, D.: Clustering by passing messages between data points. In: Science (2007)
10. Gnep, K.K., et al.: Haralick textural features on T_2-weighted MRI are associated with biochemical recurrence following radiotherapy for peripheral zone prostate cancer: impact of MRI in prostate cancer. In: J. Magn. Reson. Imaging (2017)
11. Hami, R., et al.: Predicting the Tumour response to radiation by modelling the five Rs of radiotherapy using PET images. In: J. Imaging (2023)
12. Hernández, A.I., et al.: A multiformalism and multiresolution modelling environment: application to the cardiovascular system and its regulation. In: Phil. Trans. R. Soc. A. (2009)
13. Hooker, S., et al.: A benchmark for interpretability methods in deep neural networks. In: Adv. Neural Inf. Process. Syst. (2019)
14. Joiner, M.C., van der Kogel, A.J.: Basic Clinical Radiobiology (2018). ISBN: 978-0-429-95540-2
15. Kwak, J.T., et al.: Prostate cancer: a correlative study of multiparametric MR imaging and digital histopathology. In: Radiology (2017)
16. Lundberg, S.M., Lee, S.-I.: A unified approach to interpreting model predictions. In: Advances in Neural Information Processing Systems (2017). ISBN: 978-1-5108-6096-4
17. Marinkovic, M., et al.: Comparison of different machine learning models in prediction of postirradiation recurrence in prostate carcinoma patients. In: Biomed. Res. Int. (2022)
18. Nanekaran, N.P., et al.: Prediction of prostate cancer recurrence after radiotherapy using a fused machine learning approach: utilizing radiomics from pretreatment T2W MRI images with clinical and pathological information. In: Biomed. Phys. Eng. Express. (2024)
19. Nicoló, C., et al.: Machine learning and mechanistic modeling for prediction of metastatic relapse in early-stage breast cancer. In: JCO Clin. Cancer Inform. (2020)
20. Ribeiro, M.T., Singh, S., Guestrin, C.: Why should i trust you? Explaining the predictions of any classifier. In: Proceedings of the 22nd ACM SIGKDD International Conference on Knowledge Discovery and Data Mining (2016). ISBN: 978-1-4503-4232-2
21. Samek, W., Wiegand, T., Müller, K.-R.: Explainable Artificial Intelligence: Understanding, Visualizing and Interpreting Deep Learning Models (2017). arXiv: 1708.08296
22. Sosa-Marrero, C., et al.: Towards a reduced in silico model predicting biochemical recurrence after radiotherapy in prostate cancer. In: IEEE Trans. Biomed. Eng. (2021)
23. Joost J.M., Van G., et al.: Computational Radiomics System to Decode the Radiographic Phenotype. In: Cancer Res. (2017)

24. Wang, H., et al.: Deep learning-based radiomics model from pretreatment ADC to predict biochemical recurrence in advanced prostate cancer. In: Front. Oncol. (2024)
25. Zumsteg, Z.S., et al.: Anatomic patterns of recurrence following biochemical relapse in the dose-escalation era for prostate patients undergoing external beam radiotherapy. In: Urol. J. (2015)

Author Index

A
Acosta, Oscar 152
Ahmad, Salmah 140
Akkari, Fouad Georges 47
Atad, Matan 109
Audigier, Chloé 47

B
Bartelt, Bianca 140
Bissoto, Alceu 35
Briens, Aurélien 152

C
Chan, Mark Y. 1
Chan, Mark YY 130
Cheng, Wenzheng 119
Cho, Seonaeng 69
Chourak, Hilda 152
Corrias, Alberto 79

D
De Crevoisier, Renaud 152
Dillon, Joshua 1
Ding, Hao 119
Duran, Rafael 47

E
El-Bouri, Wahbi K. 12
Enzmann, Matthias 140

F
Flühmann, Tim 35
Frings, Oliver 47

G
Gao, Yujia 23
Garcia-Tirado, Jose 35
Graf, Robert 109
Grau, Vicente 130

H
Hernandez, Rémi J. 12
Hoang, Trung-Dung 35

J
Jiang, Zichen 1

K
Kim, Ji Woong 119
Kirschke, Jan S. 109
Koch, Lisa M. 35
Kohlhammer, Jörn 140
Krieger, Axel 119

L
Lee, Kyunghyun 99
Lerchl, Tanja 109
Li, Lei 1, 90, 130
Li, Zhen 79
Lian, Xu 119
Lim, Sunghwan 99
Liu, Xiaoyue 1, 130
Lyu, Yilin 1

M
Mao, Chengzheng 23
Mauger, Charlene 1
McGinnis, Julian 109
Meister, Felix 47
Möller, Hendrik 109

N
Naik, Vihangkumar V. 35
Nash, Martyn 1
Nispel, Kati 109
Noh, Gunwoo 58

P
Paetzold, Johannes 109
Poeta, Eleonora 152

R

Rueckert, Daniel 109

S

Seo, Minjee 58, 69
Septiers, Valentin 152
Sheng, Xicheng 130
Shin, Minwoo 58, 69
Shin, Won-Yong 99
Shin, Yong-Min 99
Shlychkov, Artemii 35
Shu, Hongchao 119
Sia, Ching-Hui 1, 130
Sosa-Marrero, Carlos 152

T

Tan, Ying Zhen 23
Tonglet, Andrea 47

U

Unberath, Mathias 119

W

Wang, Shijie 90
Wang, Xinyu 119
Watrinet, Julius Maria 109
Wesarg, Stefan 140
Wolf, Ruben 140

Y

Yang, Fan 1
Yoo, Seung-Schik 58
Yoon, Kyungho 58, 69, 99
Young, Alistair 1
Yu, Chenhao 119

Z

Zhang, Yuqian 119
Zhao, Debbie 1
Zhuang, Xiahai 130
Zuluaga, Maria A. 152

Made in the USA
Monee, IL
03 May 2026

49438412R00098